AF324029

STATISTICAL
PROCESS CONTROL

QUALITY AND RELIABILITY

Edward G. Schilling

Center for Quality and Applied Statistics
Rochester Institute of Technology
Rochester, New York

Additional volumes in preparation

STATISTICAL PROCESS CONTROL

A Guide for Implementation

Roger W. Berger

Industrial Engineering Department
Iowa State University
Ames, Iowa

Thomas Hart

Eglin Air Force Base
Eglin, Florida

MARCEL DEKKER, INC.
ASQC QUALITY PRESS

New York and Basel
Milwaukee

Library of Congress Cataloging-in-Publication Data

Berger, Roger W.
  Statistical process control.

  (Quality and reliability ; 8)
  Includes index.
  1. Quality control--Statistical methods.
2. Process control--Statistical methods. I. Hart,
Thomas, [date]. II. Title. III. Series: Quality
and reliability ; vol. 8.
TS156.B466   1986   658.5'62   86-4609
ISBN 0-8247-7625-9

COPYRIGHT © 1986 by ASQC   ALL RIGHTS RESERVED

Neither this book nor any part may be reproduced or transmitted in any form or
by any means, electronic or mechanical, including photocopying, microfilming,
and recording, or by any information storage and retrieval system, without permission in writing from the publisher.

MARCEL DEKKER, INC.
270 Madison Avenue, New York, New York 10016

American Society for Quality Control
230 West Wells Street, Milwaukee, Wisconsin 53203

Current printing (last digit):
10 9 8

PRINTED IN THE UNITED STATES OF AMERICA

PREFACE

This guide is directed primarily at American business managers. However, the ideas and concepts presented will work almost anywhere. As a matter of fact, it is because statistical process control is working so well elsewhere, most notably in Japan, that American business is taking note of this "new" idea.

Statistical process control is not new; its mechanics have existed with very little refinement since first invented some 60 years ago, by an American. What is new is the worldwide competitive pressure faced by American business in such traditional American bastions as steel and automobiles. Not too many years ago the U.S. auto and steel industries, after decades of riding the crest of world leadership in quality and productivity, woke up to find not only their world markets but their domestic markets being syphoned away by foreign competition. The foreign goods were cheaper, yes, but they were also just as good and often better than their American-made counterparts.

So, what happened? How did the U.S. lead slip away? For some period after this realization that there really was competition, lots of excuses were advanced lamenting unfair foreign competition. Like most excuses they fell apart under close examination. Then American business started looking for reasons.

The examination of foreign business practices focused on Japan; they had made the most dramatic inroads into world trade. Some Japanese business practices were found which simply could not be incorporated into U.S. business, at least not in the short run. One of the most surprising discoveries, though, was that of Japan's use of statistical process control methods as one of the main pillars of their success. Here was a business practice which had been developed in the United States in the early part of this century and then exported from the U.S. to Japan in the late 1940's and 1950's. Yet these same phenomenally successful practices had never gained acceptance in the U.S. The surprise was how we let it get away from us. The Japanese used statistical methods to earn a position as a world industrial leader. U.S. industry on the other hand had turned its back on statistical process control. Discussion of statistical process control in the U.S. had been relegated almost exclusively to academia.

We are now full circle in the U.S. Not only are we re-evaluating the merits of statistical methods, we want to know how statistical process control works and how to implement it in practice.

The purpose of this guide is to strip away the mystery surrounding statistical process control and to present its concepts and principles in as simple and straightforward a manner as possible. This text will not make you an expert in developing a statistical process control program for your company, but it will certainly give you the tools you need to intelligently discuss the subject and to interpret process data. A reference listing is included at the end of the guide for further in-depth study of specific application and trouble-shooting procedures.

Roger W. Berger
Thomas Hart

CONTENTS

I. STATISTICAL PROCESS CONTROL

Statistical Process Control. A rather imposing term that means to use statistical methods to monitor the steps in a process. While the processes referred to most often are manufacturing processes, SPC is also applicable to the service industries. Any job that has a beginning, steps to be followed and an end, and costs money to perform, can benefit from statistical monitoring of the process. The goals of SPC are to improve and ensure quality and, in so doing, to minimize process costs due to waste as a result of rejects or other causes.

It is important to keep these goals in mind: to improve and ensure quality. Improving and ensuring quality throughout the steps in a process invariably reduces costs: costs of waste, costs of rejects and, significantly, costs of customer dissatisfaction with poor products. Also, it is important to note that SPC is not a cure-all for quality and production problems. It will not correct a poor product design or resolve poor employee job training. Nor will it correct an inefficient process or worn out machines and tooling. However, it will help in leading to the discovery of all of these types of problems and to identifying the type and degree of corrective action required.

BENEFITS TO BE GAINED

Traditional quality control efforts have concentrated on inspection of finished goods. Finished items are inspected by any of a variety of plans ranging from sampling of items from a production lot to a 100% inspection of all items produced. This traditional approach to quality control has two glaring deficiencies. First, no inspection plan for checking finished goods will catch all of the rejects. Even a 100% inspection of all finished goods lets some rejects through the system. Secondly, when end items are identified as rejects the damage is already done. At best, the item must be reworked at some additional cost to the system the process is a part of. At worst, the reject is an excellent example of a completely assembled piece of junk, which moves directly to the scrap pile at an even higher cost to the system.

In contrast to the traditional approach to quality control, statistical process control involves the integration of quality control into each step in the production process. Standards are established for each step and an acceptable range about each standard is determined. As long as the procedure for each step yields a product within its set range, quality is ensured. When the acceptable range is exceeded, or a trend is identified which indicates the range will soon be exceeded, the step is stopped and adjusted to bring it back in line with the standard. It should be mentioned that the statistical methods also help prevent over adjustment by indicating when the process should be left alone. In this manner quality is ensured at each step in the process and rejects are not passed along for further processing.

Statistical process control amounts to treating each processing step as the producer of a finished good, with the follow on step being the customer of that good. If standards for each step are properly set and are properly monitored and maintained, the final output of the process must meet standards by default: nothing is wrong with it. Thus, while traditional QC efforts emphasize the detection of poor quality in end items, SPC stresses the prevention of poor quality through the process. Traditional quality control efforts can be reduced and in some cases eliminated with a well designed SPC program.

Implementing SPC does involve training. It involves training management, supervisors and workers. Some companies have used formal training and some have used on-the-job training. Other companies have used a combination of both formal training and on-the-job training. The important point is that the level and type of training can be tailored to the needs of the company.

The statistical concepts involved in SPC are basic and are easily understood with minimal instruction. The mathematics involved in monitoring SPC are easily grasped by anyone with an 8th grade education.

WHO GAINS FROM SPC

Most production workers want to do a good job and are proud of their work. SPC gives them the tools they need to monitor their efforts to ensure quality, and it gives them the means to graphically illustrate problems or to prove the fruits of their efforts, enhancing both pride in workmanship and job satisfaction.

Supervisors need to learn SPC so they can ensure its proper implementation on the work floor and so they may interpret the statistical data in troubleshooting process problems. While statistical process control gives workers the proof that their work meets quality standards, it also gives supervision the key to unlock the source of where quality problems are originating.

Management needs to learn statistical methods so that they can interpret process statistical information in monitoring the output of their business. An added benefit for management from the use of SPC is that the statistical quality data can be a valuable product marketing tool. The SPC data gives management the hard figure proof of their product's quality. Many large and small companies are beginning to require proof of quality from their suppliers of everything from raw materials to semi-finished and finished goods. A properly documented statistical process control program provides all of the proof of quality required to get those contracts.

The bottom line of SPC is profits. More profits arise from increased productivity and increased productivity is what SPC is all about. Workers are given the tools to work smarter, not harder, to ensure quality with the goal of no rejects, no wasted production effort. Supervision is given the tools to coordinate the quality effort while eliminating much of the traditional antagonism between production and quality control functions. And finally, management is given the means to manage quality and reduce waste. More quality output for the same amount of input — increased productivity. That is the selling point of statistical process control. Its other benefits are also important, but they are a result of increased productivity.

A HISTORY OF SPC

Statistical process control is not new. The idea of monitoring manufacturing processes and the methods for accomplishing the monitoring process were advanced by an American statistician, Dr. Walter A. Shewhart, in the 1920's. Dr. Shewhart defined product attributes and variables, sources and types of product variation, and how to gather, plot and interpret data. As a matter of fact, with very little refinement Dr. Shewhart's methods for statistical monitoring of manufacturing processes are today's SPC methods. During the latter stages of World War II the United States's government began incorporating Dr. Shewhart's ideas into the war material manufacturing effort. However, the war came to a close before statistical manufacturing methods were firmly ingrained in the American manufacturing mind.

Following World War II the United States was virtually the only major industrial power in the world whose manufacturing organization had survived the war unscathed. The U.S. became the major supplier of manufactured goods to the non-communist world. There was essentially no real competition to U.S. dominance in manufacturing. Without the competitive pressure from abroad there was no compelling reason for U.S. firms to change old manufacturing methods. We were competing against ourselves and everyone used the same methods. Statistical methods in process control, for a variety of reasons, quickly disappeared from the American manufacturing scene. With few notable exceptions, SPC methods were absent from the North American business environment from the 1950's until the 1980's.

During this same 30 year time span, war ravaged Japan embraced the concepts of statistical process control under the tutelage of another American, Dr. W. Edwards Deming, a former colleague of Dr. Shewhart. The results of the Japanese implementing statistical methods into their manufacturing industries are legend. From a totally destroyed economy in 1945 they are now a world leader in quality and productivity.

Many reasons are given for Japan's miraculous post war recovery. Many excuses are advanced on why Japanese business practices cannot be translated into American industry. Some of those excuses have merit, but many are groundless. One of the primary pillars upon which Japanese industrial success rests is statistical process control. The facts support that SPC can be incorporated into U.S. industry, or any industrial society for that matter, with results being very quickly measured in increased profits.

SPC is an ongoing and endless program. It is a change in the way business views production. Management no longer takes a production facility and tries to drive it to its limits. Management now discovers the system's limits and works to maintain quality within those production limits, while at the same time identifying steps in the process for improvement to raise the quality and productivity limits. Again, SPC is not aimed at the detection of quality defects, it is aimed at the prevention of defects. While initial implementation of SPC can yield impressive results, the ongoing nature of the program can generate even more impressive results in the long run.

CURRENT SPC PROGRAMS IN THE U.S.

Probably the best way to sell statistical process control programs is to cite the results of some representative programs implemented in recent years in the United

States. Most of the examples presented involve using SPC in an industrial environment. With a little imagination you can see how the same concepts will easily fit into service industry process improvement and productivity improvement efforts.

The American automobile industry has recently begun to implement SPC in a big way, not only in their own production efforts but by demanding that their suppliers also implement SPC methods to ensure the quality of subcontracted automobile components. The dividends realized have been substantial.

At the Ford Motor Company Plastics, Paint and Vinyl Division/Saline Plant, SPC methods were applied to the automotive radiator grille manufacturing process. The grille manufacturing process had shown erratic quality output and efforts to stabilize and improve quality had been fruitless. Application of u-charts and Pareto diagrams to the process enabled a clear definition of the problems and helped isolate the major causes of defects. In less than 5 months the grille manufacturing process was stabilized and the average weighted defect rate per grille was reduced by 95%.

At another Ford facility, statistical methods were integrated into the manufacturing process of a fuel pump component at the Engine Division of the Essex Engine Plant. The problem with the fuel pump centered around inconsistent case hardness depth on the camshaft fuel pump eccentric causing excessive rework and scrap. The results of the collection of the statistical data and the plotting of $\bar{X}$ and R charts enabled refinement of the problem which lead to a redesign of a heating coil used in the case hardening process. Estimated cost savings for the first year of this single change in the process were $150,000.

At General Motors Corporation the Saginaw Steering Gear Division has used simple statistical methods to increase productivity and improve quality. They made improvements in quality resulting in lower costs and higher employee moral coupled with increased pride in workmanship.

The Inmont Corporation applied SPC to its production of extruded rubber weatherstripping in 1982. Inmont provides weatherstripping to both Ford and General Motors. Implementation of the use of $\bar{X}$ and R charts resulted in a more stable process with substantially reduced variation between parts; inspection time was reduced by 3 labor hours per day and reduced waste resulted in a 4% savings in material. In another auto related industry, United Technologies (Essex) used $\bar{X}$, R and p charts and realized dramatic improvements in quality and reduced variability in their production of mini-relays for automobiles. United Technologies' now high quality has earned them Ford Motor Company's highest supplier rating and a preferred supplier status.

Nashua Corporation, an early leader in the implementation of statistical process control methods, cites many examples of specific cost savings through the use of SPC. In one example, statistical methods were applied to the manufacturing process for carbonless paper in 1979. The carbonless paper production involved a paper coating process which was using too much very expensive coating material. The only solution to the problem appeared to be the purchase of a new $700,000 piece of equipment. Application of SPC to the problem showed that the problem was not the machine but the process itself. With the elimination of a few special problems, some adjustments to the process, and a little employee training material savings were estimated at $800,000 per year by the spring of 1980 — all at virtually no cost except a little training for the operator and some management time to implement SPC methods. That is a hefty return for a negligible cost.

Other examples of the successful use of statistical process control abound in recent technical and professional literature. The techniques of SPC are easy to learn with minimal training. The biggest hindrance to the wide acceptance of statistical methods is that it is new and it requires a change in the way both workers and management view production. Without question the greatest asset SPC has going for it is that it works. Properly implemented statistical methods can yield dramatic early results and will give even more substantial long term benefits.

SUMMARY

This chapter introduces the topic of statistical process control and briefly reviews its history and potential. Several contemporary examples of gains realized from the use of statistical methods in process control are cited. In the examples a number of types of charts used in SPC were mentioned and will be covered in Chapters 2 and 4. There will be a number of times in the first part of this manual where the phrase ''properly implemented'' will be used in conjunction with SPC programs. We'll look at the do's and don'ts of implementation in Chapter 5.

II. STATISTICAL CONCEPTS AND TECHNIQUES INVOLVED

Now that we have all managed to struggle through Chapter I and discussions of statistical process control in rather nebulous terms, we need to get more specific about terms and concepts of SPC to grasp how statistical methods really work. Hopefully the previous chapter at least excited you about the potential gains to be realized from the use of statistical methods in process control and piqued your interest in learning more.

Dr. W. Edwards Deming, the guru of the 1980's on the subject of statistical methods applied to process control, likes to refer to himself as an apprentice statistician. He is probably just trying to pull everyone's leg. If there is such a thing as a master statistician, he was a recognized master at least 40 years ago. His efforts in revolutionizing Japanese industrial practices have earned him a position as a master among masters in applied statistics in the realm of manufacturing. For the rest of us real neophytes in statistics, we must gain a working knowledge of the statistical terms and procedures involved in using SPC before we can intelligently discuss the subject. This chapter will address those terms and procedures.

POPULATIONS AND SAMPLES

A population is the collection of all individuals in a designated group. One might consider the population of all United States citizens. That is a finite population, at least for this instant in time. A more appropriate business example might be the population of all Monroe name brand automotive shock absorbers of a given part number. An example in the service sector would be the population of all purchase order inputs to a company's computer system.

The problem with trying to deal with and evaluate populations as a whole is that they are too large to manage. In addition, most populations in a business application are what are known as "dynamic" populations: the size of the population keeps changing — items keep coming off the production line, orders keep coming in, people are born and people die. How do you evaluate what you want to know about such populations? You use samples.

Sampling involves taking small representative groups of individuals (parts, people, paperwork, etc.) from the parent population. These samples, manageable in size, are then evaluated for whatever characteristics of the population one is interested in, and the results are translated back to estimate characteristics of the population as a whole. Sampling, properly done, will define characteristics of the population with a high degree of accuracy.

How do we know sampling works to define the population? Statistics. The intricacies of the statistical proofs of why sampling works fill volumes of statistical texts, usually found under the title of "Sampling Theory" or something similar. If you have a driving interest in learning more about why sampling works, any university book store will have a number of texts on the subject.

There are a number of ways to conduct sampling plans. Political polls are usually based on a single large sample taken from a broad and representative cross section of individuals from the population in question. From a business perspective, though, it is highly unlikely that such a broad and large sample would be of any practical interest.

For most business process control applications it is generally recommended that sample sizes should be keep small, as few as 4 or 5 per sample and not more than 20 per sample. Most samples are of size 5 to 10. Small samples taken at more frequent and regular intervals provide a highly accurate indication of process performance. SPC is geared to the taking of small samples at regular intervals.

One small note concerning terminology is important at this point, especially if you ever find yourself talking statistics with a statistician. Characteristics of parent populations are known in statistical jargon as parameters. For instance, let's say we have a warehouse population of half inch steel rod and we measure every rod to find its exact diameter. We find, after spending the better part of a lifetime measuring, that the actual average rod diameter is 0.505 inches with actual sizes ranging from 0.485 to 0.526 inches. That average rod diameter value of 0.505 inches is a population parameter. We could have achieved the same results by taking random samples of the stock in the warehouse and done it all in a half a day or so. But, the results of the sampling plan would have been only estimates of the population parameter. That is the distinction: only populations have parameters, sampling plans yield estimates of parameters. Unless you have measured every item in a population a statistician will argue that you do not know the population parameter with absolute certainty. Your sampling effort has only given you an estimate of the parameter, however accurate. It is a fine line distiction, but an important one in keeping terminology straight.

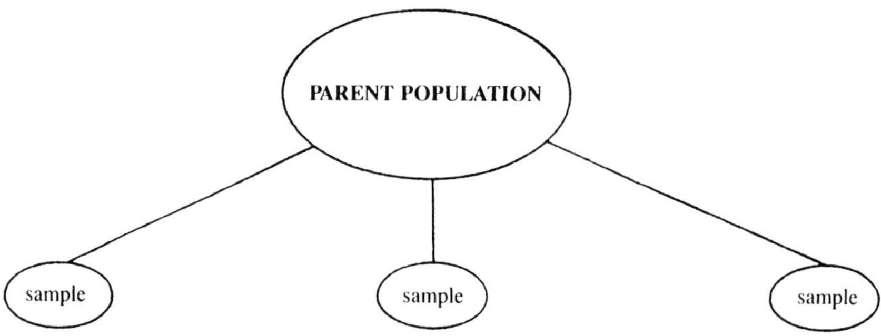

From a parent population representative samples are drawn and manipulated. From these samples estimates are computed and from those estimates inferences can be drawn concerning population parameters.

MEAN, MEDIAN, and MODE

Most of us know more statistics than we give ourselves credit for. It's the jargon that keeps us in the dark when it comes to discussing or using statistics. That fact is no more true than when discussing the terms mean, median and mode. Each is a measure of central tendency in the parent population.

The mean of a sample is generally designated by the term "$\overline{X}$" (called X bar). The mean is nothing more than the average value of the measurements taken on a sample. For example, referring back to our warehouse full of ½ inch steel rod, let's say we take a sample of five rods and our interest is again in the diameter of the rods. We measure each of the five rods and come up with diameters of: 0.498, 0.510, 0.501, 0.507, 0.506 inches respectively. To find the average rod diameter of the sample we add up the individual measurements and divide by the number of individuals in the sample. Thus, the average diameter — mean diameter = $\overline{X}$ = 0.5044 inches. Carrying this one step farther, SPC involves taking many samples over regular and continuing intervals. Each sample has an $\overline{X}$. Adding the $\overline{X}$'s and dividing the sum by the number of samples taken yields $\overline{\overline{X}}$ (called X double bar), the average of averages or the mean of the sample means. This $\overline{\overline{X}}$ is an estimate of the average of the population as a whole.

The median of a sample is the middle value of the sample. It is easiest to find in an odd size sample. Take our previous sample of 5 rods and arrange the measured diameters in ascending order: 0.498, 0.501, 0.506, 0.507, and 0.510 inches. The middle value is 0.506 inches, and that is the median for this sample. For an even sized sample the computation of the median is only slightly more complicated. After arranging the measured values in ascending order take the two middle values, add them together and divide by 2. Simple.

The last measure of central tendency is the mode. The mode is the most frequently observed observation or measurement in a sample. In our example of the five rod sample there is no mode, each measurement occurs only once. In a different sample where a given measurement occurs more than once, the most frequently observed measurement is the mode.

RANGE

The range is a measure of the difference in measurement between the largest individual measurement taken in a sample and the smallest individual measurement. From our example of the 5 rod sample the rod diameters were: 0.498, 0.501, 0.506, 0.507, and 0.510 inches. The range of the sample is the difference between the largest and smallest values, or 0.510 − 0.498 = 0.012 inches. As in the case of determining $\overline{\overline{X}}$, ranges are computed for many samples, their values are summed and divided by the number of samples involved. This process yields an average range, normally denoted by the term $\overline{R}$.

While finding $\overline{X}$'s on individual samples and then computing $\overline{\overline{X}}$ on the group of samples will give you a measure of what a process turns out on average, computing sample R's and $\overline{R}$ will give you a feel for how widely measurements differ between individuals, in other words, how stable is the process. We will talk more about stability later. In the mean time, in relating the computations of $\overline{\overline{X}}$ and $\overline{R}$ to a real life process, obviously the goal is to have $\overline{\overline{X}}$ located somewhere within acceptable standards with as narrow a variation between individuals as necessary to keep all individuals within acceptable limits. We will also address more on how you work towards this goal in Chapters 3 and 4.

To get a feel for the simplicity of the computations involved in the determination of the mean ($\overline{X}$), mean of the sample means ($\overline{\overline{X}}$), median, mode, range (R) and average range ($\bar{R}$), a short example is in order. Consider 3 samples and assume that the measurements have been rearranged in ascending order for ease of determining the median and mode.

Sample 1	Sample 2	Sample 3
7.35	7.37	7.36
7.40	7.38	7.38
7.42	7.44	7.41
7.43	7.45	7.43
7.44	7.46	7.44
7.44	7.46	7.45
7.46	7.48	
Sum = 51.94	52.04	44.47
Mean = 51.94/7	= 52.04/7	= 44.47/6
= 7.42	= 7.432	= 7.412
Median = 7.43	= 7.45	= (7.41 + 7.43)/2
		= 7.42
Mode = 7.44	= 7.46	= undefined
Range = 7.46 − 7.35	= 7.48 − 7.37	= 7.45 − 7.36
= 0.11	= 0.11	= 0.09

Mean of Sample Means ($\bar{\bar{X}}$) = (7.42 + 7.432 + 7.412)/3
$$= 7.422$$

Average Range ($\bar{R}$) = (0.11 + 0.11 + 0.09)/3
$$= 0.103$$

Any and all computations of these factors, $\bar{X}$, $\bar{\bar{X}}$, R, $\bar{R}$, median and mode, always follow these same computational steps. There may be a little more adding and a little more subtracting depending on the size and number of samples, but the computational steps are always the same. Elementary arithmetic.

VARIANCE AND STANDARD DEVIATION

At this point in our discussion of statistical concepts there are only three more points to cover: variance, standard deviation and distribution patterns. It is difficult to grasp variance and standard deviation without understanding distribution patterns, and distribution patterns are of limited use without the variance and standard deviation. As a result, you may have to jump back and forth between the two sections for reference.

Variance is a measure of the difference of individual observations from the average (mean) value. You will seldom see the term used in relation to statistical process control. The computations involved in finding the variance, while not particularly complicated, are tedious and we will leave the subject without investigation. It should be noted, though, that variance measurements are used when making comparisons between the variableness of two similar processing lines making the same or similar parts. If you have an interest in this area of variance comparisons between similar and parallel processes, several of the texts noted in the reference section of the appendix have good presentations on the topic and examples of the equations and computations involved.

The measure of variability used most often in SPC is the standard deviation, denoted by σ (sigma). The standard deviation is the square root of the variance. It yields a measure of population variability which can then help estimate the percent of the parent population which falls within a given number of standard deviations from the mean (average).

For the astute reader, having noted that we bypassed the computation for determining the variance, you may wonder how we are going to arrive at the square root of a value we did not calculate. Lucky for you, mathematicians and statisticians have been working with sampling efforts for a long time and have developed some short cuts.

For a stable normally distributed parent population an estimate of the process standard deviation, $\hat{\sigma} =$ (called "sigma hat"), can be calculated from $\bar{R}$, the average of sample ranges. The equation to use is

$$\hat{\sigma} = \bar{R} / d_2,$$

where d_2 is a constant dependent on sample size.

Values for d_2 can be found in Table A2 in the back of this guide. From your sampling effort you would know the number of items taken per sample (the sample size) from the measurements you can calculate $\bar{R}$. Once you have the sample size and $\bar{R}$, look up the sample size on the d_2 table, punch in the numbers on your trusty calculator, and you have $\hat{\sigma}$, an accurate and adequate estimate of the standard deviation, σ.

There is a long hand expression for calculating the standard deviation which is occasionally used in SPC. We will look at it in the discussion on "s" charts. For most applications the shorthand route to finding $\hat{\sigma}$ is entirely adequate. The significance of the standard deviation in defining the population will become apparent after we define distribution patterns.

DISTRIBUTION PATTERNS

There are five different distribution patterns that are commonly referenced in business applications. The patterns are the normal, log normal, binomial, Poisson and exponential.

The exponential distribution is used to model the interval between events occurring at random. An example is the time between process shut-downs, applied to a process which is shut down at random times for adjustment, repair, restocking of material, and so forth. The Poisson distribution has several different quality control applications. It is best known as a means of approximating the binomial distribution, and this approximation is often used to design sampling plans. It is also the basic underlying distribution for the "c" chart, which is discussed on page 47. The Poisson distribution describes the number of occurrences of random events per fixed interval. The number of nicks per yard of wire, the number of pinholes per sheet of newsprint, and the number of arrivals at a tool crib per hour, are examples of Poisson variables.

The next distribution pattern we will look at is the binomial distribution.

Bi- means two; in this case it means the result of an inspection can only have one of two possible outcomes. The classic example is flipping a coin that can only come up heads or tails. Assuming the coin is fair, if we flip the coin fifty times we would expect the coin to come up heads 50% of the time and tails 50% of the time. However, in any one sample of 50 coin tosses the number of times a coin turns up heads may be more or less than 25. For many samples of 50 coin flips we would expect the mean probability of a fair coin turning up heads to be 50% of the time, or 25 times out of 50. Additionally, we would expect the probability of the coin turning up heads more or less often than 25 times in 50 to become progressively less and less as we moved farther and farther away from this mean value of 25 toward the extremes of 0 and 50 for any one sample.

The binomial distribution is directly applicable to SPC when we apply statistical methods to product attributes. Attributes will be defined and discussed in detail in Chapter 4. At this point the important thing to remember about the binomial distribution is that when dealing with relatively large samples (generally, 50 or more), which is the case when applying statistical methods to attribute evaluation, as long as it is expected that the less probable of the two possible outcomes will occur at least 4 or 5 times in each sample then the binomial distribution over many samples is very closely approximated by the normal distribution.

Our fourth distribution pattern is the log normal distribution. Log is short for logarithm. A log normal distribution infers that if you take the logarithm of each of the individual points or measurements in a log normal distribution and then replot these numbers you will have transformed the distribution to a normal distribution (hence its name). The log normal distribution has specific applications in process control and will be reintroduced and and discussed later in this chapter in conjunction with the discussion on R charts.

The distributions discussed above are all useful in their own special ways. However we have saved the most useful and the most important distribution for the last: the normal distribution. Let us now focus our attention on this most elegant model.

The reason the normal pattern is used is two fold. First, the normal pattern probably gives a good description of all populations that you will be concerned with in the realm of process control. Second, there is a statistical concept called the "Central Limit Theorem". This theorem tells us that when we take many samples from a parent population the means of the individual samples ($\bar{X}$'s) will be approximately normally distributed, regardless of the distribution of the parent population. This second reason works out nicely for SPC efforts, even if a parent population is not quite normal it will appear so through the sampling program. The only time this centering of the population through the affects of the "Central Limit Theorem" leads to problems in SPC is when we are dealing with a log normal distribution; it can give us a distorted view of reality in this case. None-the-less, the problem is easily recognized and avoided, as we shall see.

A description of the normal distribution is now in order. The clearest way to describe a normal distribution is with a picture.

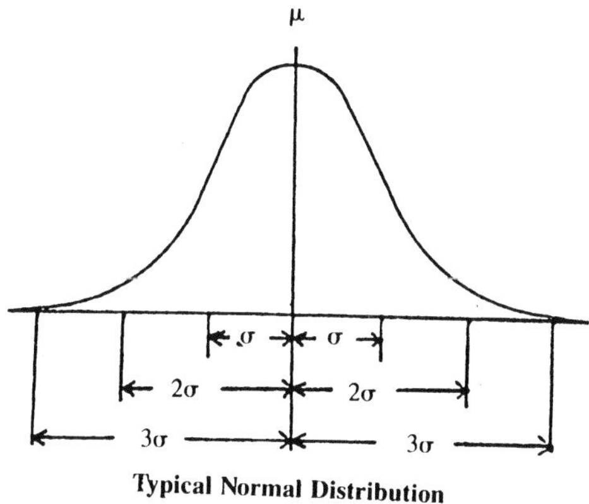

Typical Normal Distribution

Mu (μ) is the center of the distribution, its mean. We estimate Mu through sampling by calculating X double bar ($\overline{\overline{X}}$). For a normal distribution the mean, median and mode all define Mu, they are one and the same. All three were defined earlier because in a sample they might not all be the same.

The diagram also shows how the standard deviation fits into the distribution pattern. For a normal distribution pattern the distribution is symmetrical. In other words, from your population mean, Mu, there are as many items above the mean as there are below it. From our previous example of the warehouse full of steel rods our mean diameter was 0.505 inches. If the population was normally distributed this would mean that there were as many rods with a diameter greater than 0.505 as there were with diameters less than 0.505. The diameters of most of the rods would be expected to be close to the mean diameter, with fewer and fewer rods to be found at increasing differences from the mean. As can be seen from the diagram, as you move away either direction from the mean (Mu) the line defining the distribution slopes down. This illustrates the declining probability of an event occurring as you move farther away from the mean. From our steel rod example, actual rod sizes varied from 0.485 inches to 0.526 inches. If the population were normally distributed this would tell us that the probability of picking a single rod out of the warehouse of either 0.485 or 0.526 inches would be almost nil, with the probability increasing progressively as we moved in from either measurement extreme toward a peak at the mean value of 0.505.

We use the standard deviation to evaluate the probability of being within certain limits of our population distribution. Once the standard deviation (sigma or σ) is defined and the mean is defined, the population is defined. Statistical tables reference the probabilities of being at certain relative distances from the mean as measured in standard deviations. For example, for a stable population slightly more than 68% of the population can be expected to be within the range of -1 standard deviation and $+1$ standard deviation from the mean (± 1 sigma or $\pm 1 \sigma$). Likewise, slightly more than 95% of the population lies within ± 2 sigma from the mean and slightly more than 99.7% of the population lies within ± 3 sigma. All three of these percentages are commonly used and are usually rounded to 68%, 95% and 99.7% respectively.

The statistical tables used in referencing these sigma values are called "Z" tables. There is a "Z" table in the appendix of this guide. The Z value is simply a variable which stands for the number of standard deviations from the mean. Theoretically Z values can range from plus to minus infinity. From a practical standpoint, from plus to minus 3 σ (plus or minus 3 standard deviations from the mean) you encompass 99.7% of the population. This means that, by pure chance alone, 3 items in 1000 will lie outside the range of plus to minus 3 sigma.

The way the Z chart is read is to pick a Z value to the first decimal place and find it along the left hand column of the chart. Then, going along the top of the chart choose the second decimal place for the Z value. At the intersection of the row and column chosen you will find a 4 place decimal fraction. This fraction tells you the percent of the population which lies to the left of that Z value.

As an example, let's say we wish to know the percent of the population which is less than + 1.75 standard deviations from the mean. You would find 1.7 in the left hand column of the table (the Z column) and then going across the top of the table find 0.05. At the intersection of this row and column is the decimal fraction 0.9599. This means 95.99% of the population lies to the left of 1.75 standard deviations from the mean.

You must be careful here about what you are looking for. If you will notice, the example looked at the percentage of the population less than + 1.75 σ from the mean. We did not ask for the percent of the population less than 1.75 σ from the mean. If we had, then we would have been looking for the percent of the population between − 1.75 and + 1.75 Z, and that is different. In the case where you are looking for the percent of the population between two Z values, which for SPC is the normal case, you find the percent of the population for the larger Z value and subtract the percent of the population for the smaller Z value. To find the percent of the population between ±2 standard deviations from the mean look up the tabled value for + 2.00 σ and subtract the tabled value for − 2.00 σ : 0.9773 − 0.0228 = 0.9545. At these relatively large values of Z the left hand tail from the population distribution may not appear overly significant, at smaller and smaller Z values the differences become dramatic. So don't forget to subtract when you are looking for population percentages between two values.

STABILITY

So now you say, "OK, I've got the normal distribution down pat : mean, standard deviation, Z values. You've seen one you've seen 'em all." Not exactly. One of the best illustrations of this is the set of illustrations on page 16 which were taken from a Ford Motor Company training manual. The normal distribution is characterized by a bell shaped curve. In practice, the base line of the curve would be scaled reflecting appropriate units of measurement. If the population is narrowly distributed (it has a narrow range of variation between measurements) the curve has a high peak and narrow base. If the control of the process is rather sloppy the curve is flattened out with a low peak and wide base (reflecting a wide range of variation between parts). The 99.7% of the population of each curve still lies between plus and minus 3 sigma from the mean, it is just that the value of the sigma (in base line measurement units) is different for each curve.

And then there is the problem of the curves moving on you. For one reason or another something in the process can shift: a machine adjustment slips, a sprayer gets clogged, a new person on the job, a new order clerk goes to work, any of a number of things. The result could be a shift in the location of the distribution in relation to the based line.

Also, the output of the process may suddenly become erratic, a so called "out of control" condition, for any of the same reasons that could result in a population shift plus a few more. An out of control condition (let's say the bushings wear out on a motorized tool allowing it to wobble) can not only shift a population distribution, it can flatten it out, increasing its spread in a very short period of time. If several process problems are working together against the control of the process you can have a difficult time defining what the process is capable of at all.

When using statistical methods for process control a process must first be stable and in control. If it is not, trying to force statistical methods on sample data will yield conflicting and confounding results. Fortunately, SPC will quickly indicate an unstable process. Causes of the lack of stability must be isolated and corrected before a continuing SPC program can be applied for long term advantage.

SPC, then, is a plan for using statistical concepts to determine first if a process is in control and then to keep it in control while working to narrow variability to an optimal level. You don't have to draw little bell shaped curves and determine 1, 2 and 3 sigma locations each time you use SPC, but conceptually you should understand what a normal distribution is. Through the use of a couple of simple charts, and variations of those charts, SPC embodies all of the above statistical concepts into easy to use management tools. Tools that, as an old friend used to say, "normal people" can use.

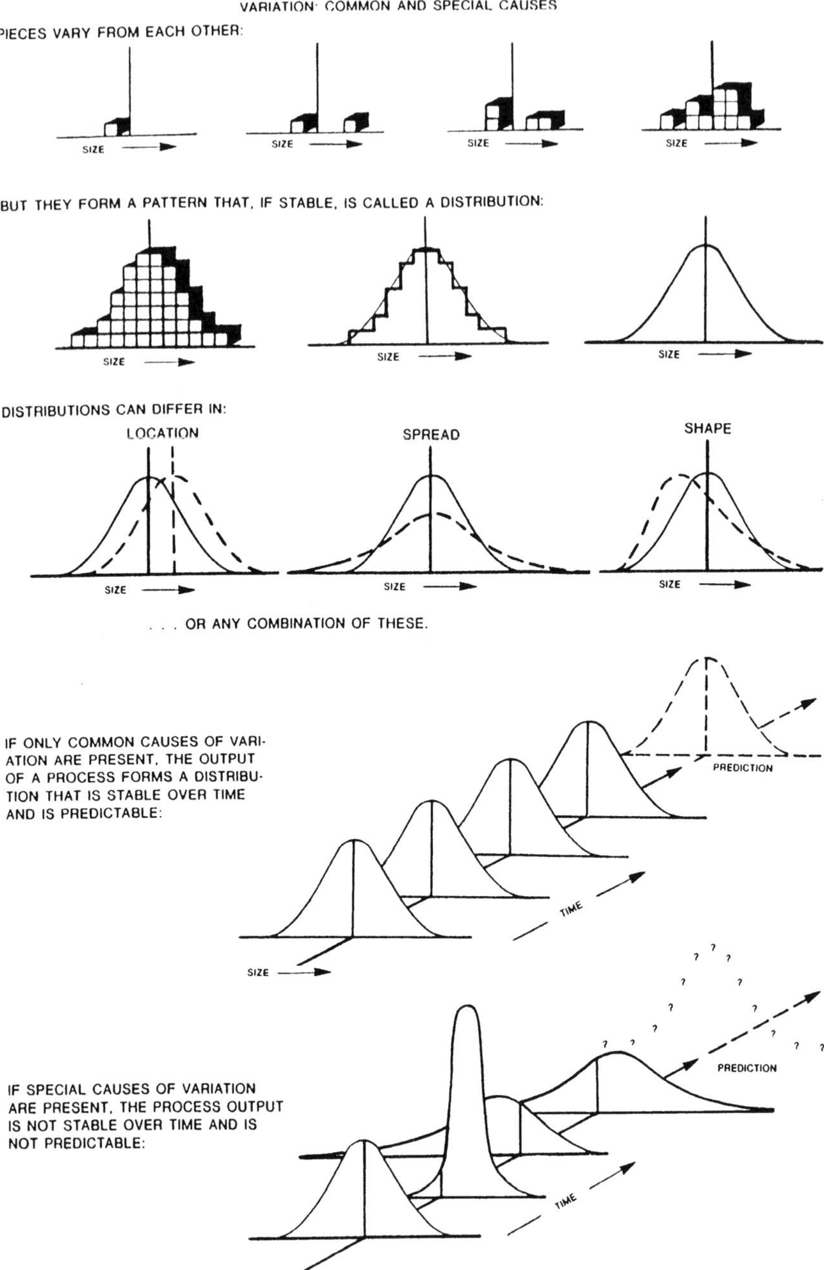

Reprinted with permission from *Continuing Process Control and Process Capability Improvement. A Guide to the Use of Control Charts for Improving Quality and Productivity for Company, Supplier and Dealer Activities.*
Ford Motor Company Booklet No. 80-01-251, July 1983

CONTROL CHARTS

The body and soul of statistical process control is contained in two charts: the $\bar{X}$ and R charts. Some consultants might argue that a number of other charts are useful and necessary. True. We will look at other charts also, but they are all variations of the $\bar{X}$ and R charts adapted to a more convenient format for specific applications.

These two charts, the $\bar{X}$ and R charts, are worked together, and the first thing you need is some data. If you are going to implement SPC on some process, spend a little time planning in an effort to decide what you want to measure and how to do the measuring. However, don't spend too much time planning; planning can become an end in itself and you need data now to get started. Decide on a sample size and a period of time between samples and start collecting samples.

When starting a process control program remember, sample sizes of 4 or 5 are adequate and sizes greater than 10 are probably unnecessary and overly time consuming. As for a time interval between samples, when you are just starting you may want samples every ½ hour, or every hour or 2 hours. Whatever interval you feel will give you a running fix on what the process is generating for output is fine, just be consistent. After the SPC program is in place and going, an interval of 3 or 4 hours or more may be quite adequate to signal process problems. But to start you need to collect the data more frequently so you can build your data base and begin controlling the process.

The best way to explain how to gather and use the data for charting is through an example chart. The following chart is from a Ford Motor Company training guide on SPC.

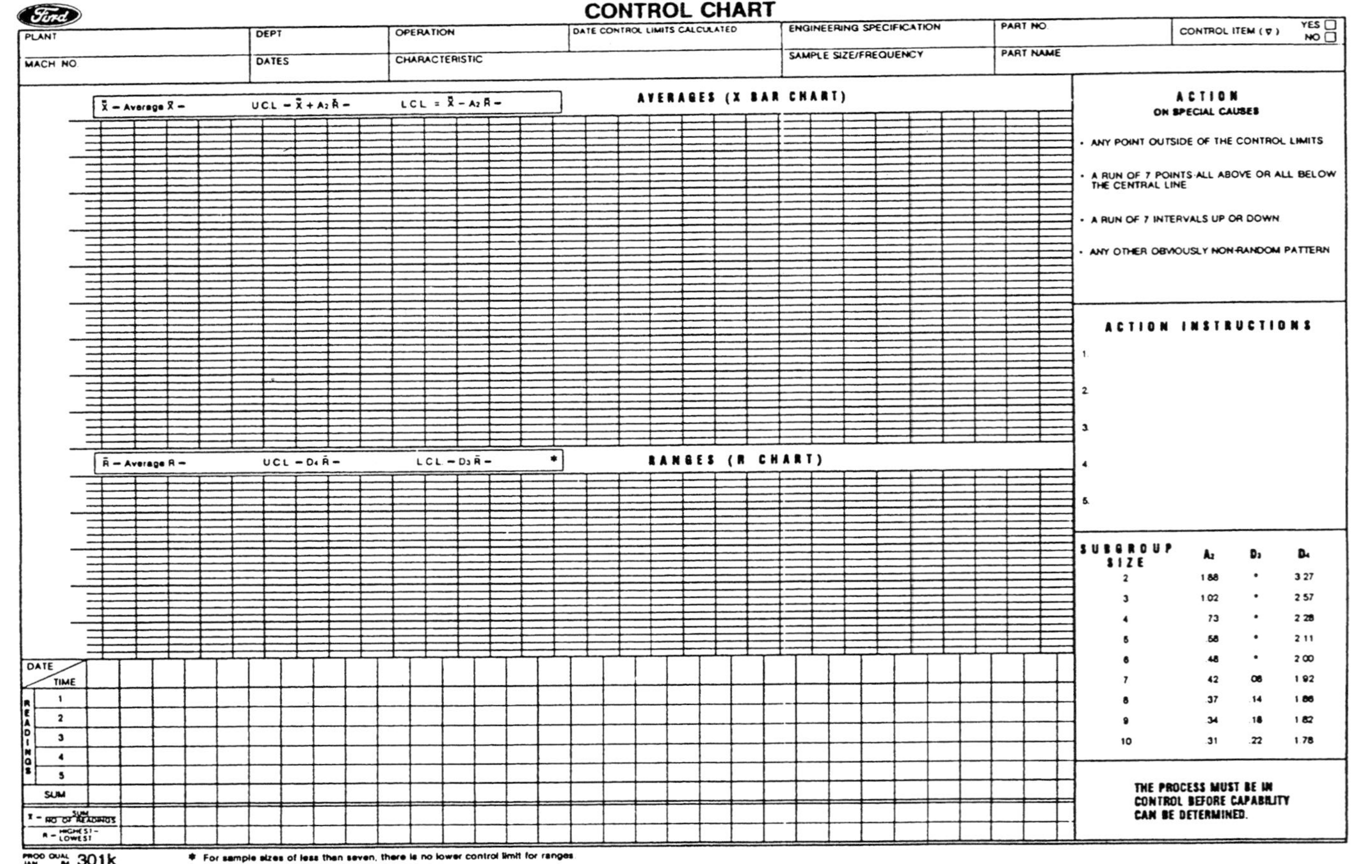

Reprinted with permission from *Continuing Process Control and Process Capability Improvement. A Guide to the Use of Control Charts for Improving Quality and Productivity for Company, Supplier and Dealer Activities.*
Ford Motor Company Booklet No. 80-01-251, July 1983

18

CONTROL CHART

PROCESS LOG SHEET

ANY **CHANGE** IN PEOPLE, MATERIALS, ENVIRONMENT, METHODS OR MACHINES SHOULD BE NOTED. THESE NOTES WILL HELP YOU TO TAKE CORRECTIVE ACTION WHEN SIGNALED BY THE CONTROL CHART.

DATE	TIME	COMMENTS

DATE	TIME	COMMENTS

Reprinted with permission from *Continuing Process Control and Process Capability Improvement. A Guide to the Use of Control Charts for Improving Quality and Productivity for Company, Supplier and Dealer Activities.*
Ford Motor Company Booklet No. 80-01-251, July 1983

As you can see, Ford has designed a convenient control chart that encompasses both the X bar ($\bar{X}$) and R chart on the same form. The top of the chart describes the "who, what, when and where" of what is being measured. At the bottom of the chart is a place to record the date and time each sample was taken, followed by a place to record the measurements (readings) for each item in each sample. This chart is designed for sample sizes up to 5. There are appropriate equations annotated and blocks to record the results of computations. Pertinent notes are included on the right hand side of the chart along with needed constants for the computations (We will look at the action notes and their significance later in Chapter 4).

On the reverse side of the chart is a place for the sampler to make notes on anything of consequence that changes in the process during or between samples. These notes are extremely important once the charting process begins. If points on the chart show sudden problems or shifts, a look at these comments will quite often quickly indicate the cause of the shift, i.e., a new man on the job, machine was adjusted, broken drill bit, etc.

Ford designed this chart to fit its SPC program and its needs. It is convenient for workers to use. Everything is clearly labeled. Symbols, equations and constants are right on the chart so workers do not have to rely on memory or have to look them up every time they need them: they have it all in front of them. Appropriate notes and comments are printed on the chart as memory joggers for the users. It is a well thought out and well organized means for recording needed data and important notes. Any company preparing to embark on SPC would be well advised to develop a similar chart designed to meet their specific program needs. If a chart is not clear recording errors will be common, rendering the data useless. If the chart is difficult or frustrating to use, it won't be used, or at best it will be used improperly.

Once your chart for recording data is ready, to get a good feel fc how the process is performing at the start a preliminary set of samples needs to e taken. 20 to 30 samples over a representative time period are necessary, probably ncompassing several days and all shifts. Several samples during each shift need to be taken to see if there is a change in output as each shift progresses. Readings should come from all shifts to evaluate if process output varies between shifts. Finally, output needs to be checked over several days to ensure the quality of the process output is consistent over time.

After you have taken your samples and recorded the sample measurements, if your chart is well designed, the data will be conveniently organized in a time ordered sequence and you are ready to manipulate the data to have it tell you about the process. First, calculate $\bar{X}$ for each sample by summing the sample readings and dividing by the number of items in each sample. Next, calculate the range (R) of readings for each sample by subtracting the smallest sample reading from the largest sample reading. (Actually, you would probably calculate sample $\bar{X}$ and R values as you took the samples and then plotted the points at the same time, but for clarity at this point we will assume all calculations and plotting are done after the final initial sample is taken).

Once all of the $\bar{X}$'s and R's for the initial samples are taken $\bar{\bar{X}}$ and $\bar{R}$ need to be calculated. As you will remember, $\bar{\bar{X}}$ is found by summing all of the $\bar{X}$ values and dividing by the number of samples used. Similarly, $\bar{R}$ is found by summing all of the R's and dividing by the number of samples used.

If you will recall back to the discussion of standard deviations, the area between ± 3 standard deviations from the mean (the mean is estimated by $\bar{\bar{X}}$) encompasses 99.7% of a normally distributed population. If we calculate the value of plus and minus 3 standard deviations from the mean and establish these lines as measurement limits, we should expect virtually all of our $\bar{X}$ points to fall within these limits.

As we mentioned before, for small sample sizes there are handy shortcut formulas for calculating the 3 sigma limits for both $\bar{X}$ and R charts (these shortcuts are another good reason for keeping sample sizes small). The plus 3 sigma limit for the X bar chart is called its Upper Control Limit (UCL) and is found by the formula:

$$UCL = \bar{\bar{X}} + A_2\bar{R}$$

The minus 3 sigma limit for the X bar chart is called its Lower Control Limit (LCL) and is found by the formula:

$$LCL = \bar{\bar{X}} - A_2\bar{R}$$

The values for the constant A_2 are dependent on the size of the sample (noted as the subgroup size on the Ford chart form). A_2 values as well as other constants can be found in the appendix.

The calculations of control limits for the R chart is slightly different than calculating control limits for the X bar chart. The sample X bar's from a stable population will be normally distributed, the R's will not. The distribution of sample ranges approximates a log normal distribution, which was introduced earlier in this chapter. The lower limit of the range of measurements between items in a given sample is zero, i.e., no difference in measurements between parts is the lower limit of the difference between parts. Further, for a stable normally distributed population we would expect the range of measurements for most samples to be located toward this lower end. Larger and larger sample ranges would occur less and less frequently. This yields a distribution of ranges which is skewed to the right, as illustrated below.

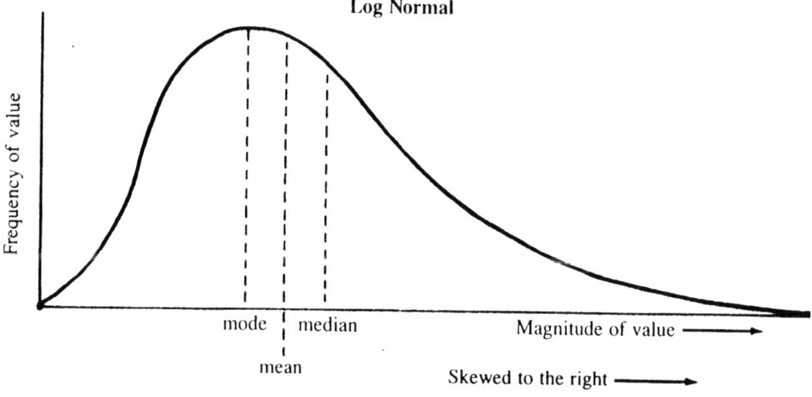

As you can see, the mode (most frequent value) is to the left of the mean (average value) which is, in turn, to the left of the median (middle value). Also, as a note of caution, remembering that a log normal distribution can be transformed into a normal distribution by taking the logarithm of each point on the log normal distribution and replotting them, a log normal distribution transformed to a normal distribution does not yield a normal distribution which indefinitely tapers off in either direction; the left hand tail is truncated at zero, the absolute lower limit of a log normal distribution.

So, how do these intricacies of the log normal distribution impact the plotting of the R chart? They mean nothing more than that two constants are needed to calculate the upper and lower control limits for the R chart. The appropriate control limit formulas are:

$$UCL = D_4\bar{R}$$
$$LCL = D_3\bar{R}$$

However, to give us all a break from the number crunching, if sample sizes are kept to no more than seven there is no lower control limit to calculate; it goes to zero by default.

Once you have all of this information from your sampling effort, the X bar's, R's, X double bar with upper and lower control limits and R bar with upper and lower control limits, it is time to plot the data to learn what there is to learn from the process. Appropriate graphs need to be set up, one each for plotting $\bar{X}$ values and for plotting R values. Appropriate graph scales need to be established for each graph so all calculated data points fall within the graph area. On the X bar chart the centerline of the distribution, as estimated by X double bar, should be highlighted, as should the lines indicating both the upper and lower control limits. Similarly on the R chart the lines indicating $\bar{R}$ and the upper and lower control limits should be highlighted. Next, if you have not already done so, each individual $\bar{X}$ and R value should be entered on the appropriate chart *in the sequence order in which it occurred.*

Looking back at the Ford chart form you see they have the graphs for both the X bar and R chart on the same form. All the user has to establish is the appropriate scale for the chart. Both graphs lie directly above the section for annotating sample readings and the vertical lines on the graphs are centered above each column for recording sample readings. This form makes both sample recording and plotting convenient for the user. After each sample is taken the appropriate measurements can be gaged and recorded. The $\bar{X}$ and R for each sample can then be calculated, recorded and plotted — all in the same vertical column.

Once the data points are plotted and the averages ($\bar{\bar{X}}$ and $\bar{R}$) and their corresponding control limit lines are drawn, it is time to evaluate the reasons for the positioning of the individual data points.

RANDOM AND ASSIGNABLE CAUSES

There are reasons for the positioning of every point on the $\bar{X}$ and R charts. In statistical jargon these reasons are called ''causes''. There are two types of causes: random and assignable. If the data points lie within the control limits established, their variation in positioning is probably due to what are known as ''random causes''. No process will produce all items on the mean: there will be some float, so to speak, about this center value. This float is called inherent process variability, you cannot get rid of it. Therefore, the inherent process variability accounts for most of the random causes.

This is not to say that as long as all points remain within the control limits there are no problems. You have to look at the charts as a whole to use them. On the X bar chart most points should be reasonably in toward the $\overline{\overline{X}}$ line with approximately the same number of points on either side of the line, not exactly the same number on each side but in the ballpark. The same applies to the R chart. If data points jump from one control limit extreme to the other but lie within limits, there is still a problem. Something in the process is unstable and needs to be investigated. If a series of succeeding points shows a trend toward a control limit, you do not have to wait until a control limit is reached or exceeded to take action. Early corrective action based on trends may well save considerable time and expense.

Even before looking at your plotted data points for the possibilities of trends and strange data patterns within control limits, you should easily be able to spot any data points which fall outside the control limits. For points that fall outside of the control limits there is almost absolute surety of an identifiable concrete reason for the outage. These identifiable reasons are called ''assignable causes'' of variation.

Again, 99.7% of a normally distributed population will lie within the range of plus to minus 3 sigma from its mean. This means two things. First, by pure chance alone (random causes) a data point will fall outside the 3 sigma control limits approximately 3 times in 1000. Second, if we turn the numbers around a bit, if a data point falls outside these control limits and you pack your guns and call out the troops to go find the problem, you will only be wrong and wasting everyone's time, on average, 3 times out of 1000. *That* is a pretty darned good decision making success rate.

We will look more at chart interpretation and expand on these few brief ideas seen here in Chapter 4.

SUMMARY

In summary, the X bar and R charts are important management tools for controlling processes, for identifying problems and for spotting trends before they become problems. This was a long chapter. It covered the majority of terms used in SPC and introduced you to SPC's two most powerful tools in the X bar and R charts. The next page shows one of the Ford Motor Company charts with sample measurements filled in and the $\overline{X}$ and R charts plotted to give you an idea of what one of these charts might look like in practice.

Before going on to look at using and interpreting charts, the next chapter will look at determining first if the process is in control.

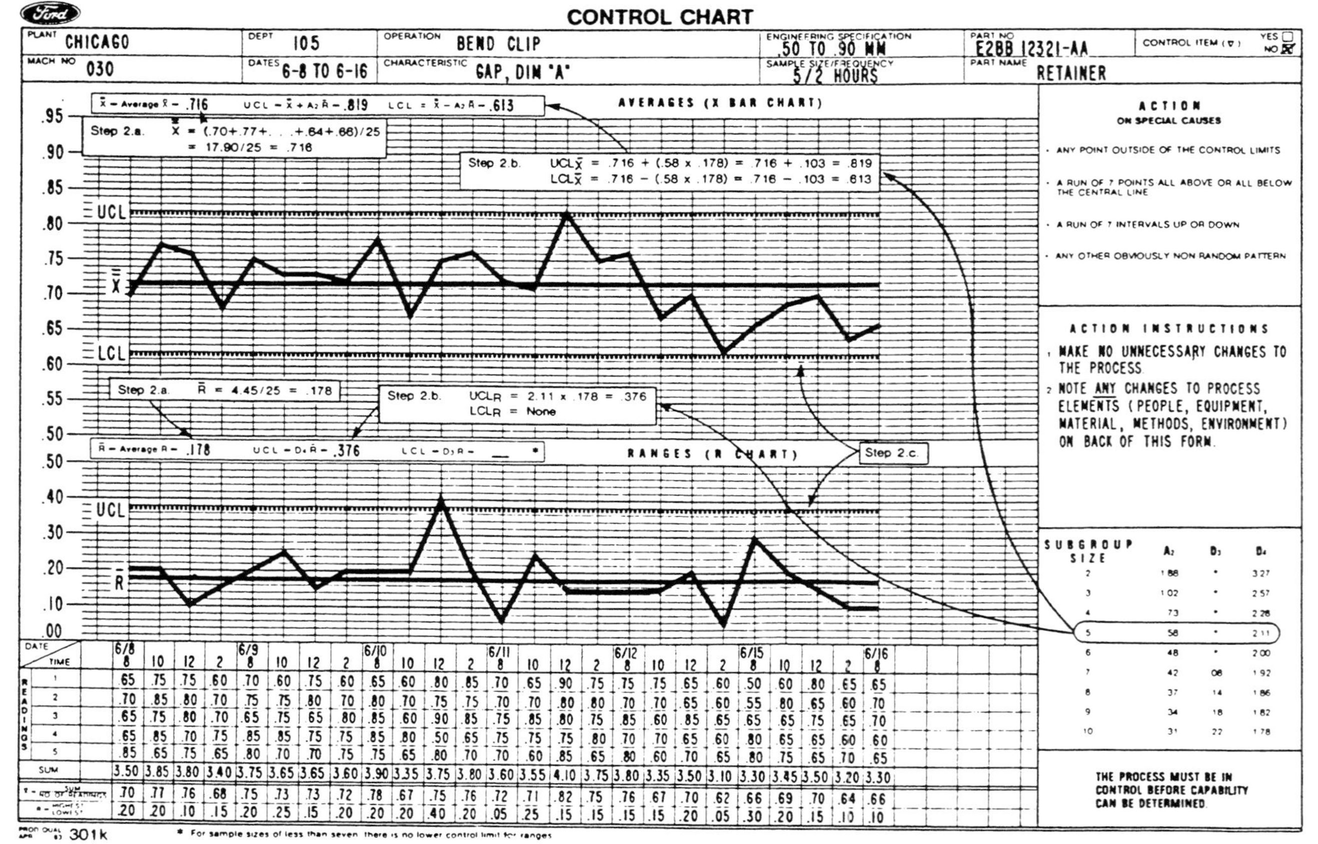

SUBGROUP SIZE	A₂	D₃	D₄
2	1.88	*	3.27
3	1.02	*	2.57
4	.73	*	2.28
5	.58	*	2.11
6	.48	*	2.00
7	.42	.08	1.92
8	.37	.14	1.86
9	.34	.18	1.82
10	.31	.22	1.78

DATE	6/8				6/9				6/10				6/11				6/12				6/15				6/16
TIME	8	10	12	2	8	10	12	2	8	10	12	2	8	10	12	2	8	10	12	2	8	10	12	2	8
READINGS 1	.65	.75	.75	.60	.70	.60	.75	.60	.65	.60	.80	.85	.70	.65	.90	.75	.75	.75	.65	.60	.50	.60	.80	.65	.65
2	.70	.85	.80	.70	.75	.75	.80	.70	.80	.70	.75	.75	.70	.70	.80	.80	.70	.70	.65	.60	.55	.80	.65	.60	.70
3	.65	.75	.80	.70	.65	.75	.65	.80	.85	.60	.90	.85	.75	.85	.80	.75	.85	.60	.85	.65	.65	.65	.75	.65	.70
4	.65	.85	.70	.75	.85	.85	.75	.75	.85	.80	.50	.65	.75	.75	.75	.80	.70	.70	.65	.60	.80	.65	.65	.60	.60
5	.85	.65	.75	.65	.80	.70	.70	.75	.75	.65	.80	.70	.70	.60	.85	.65	.80	.60	.70	.65	.80	.75	.65	.70	.65
SUM	3.50	3.85	3.80	3.40	3.75	3.65	3.65	3.60	3.90	3.35	3.75	3.80	3.60	3.55	4.10	3.75	3.80	3.35	3.50	3.10	3.30	3.45	3.50	3.20	3.30
X̄ = SUM/NO. OF READINGS	.70	.77	.76	.68	.75	.73	.73	.72	.78	.67	.75	.76	.72	.71	.82	.75	.76	.67	.70	.62	.66	.69	.70	.64	.66
R = HIGHEST - LOWEST	.20	.20	.10	.15	.20	.25	.15	.20	.20	.20	.10	.20	.05	.25	.15	.15	.15	.15	.20	.05	.30	.20	.15	.10	.10

Reprinted with permission from *Continuing Process Control and Process Capability Improvement.
A Guide to the Use of Control Charts for Improving Quality and Productivity for Company, Supplier
and Dealer Activities.*
Ford Motor Company Booklet No. 80-01-251, July 1983

III. PROCESS CAPABILITY

Before we go any further on discussing how to use and interpret charts in an ongoing SPC program we first need to look at the subject of, "Is the process capable of meeting standards to start with?" If you do not first ascertain that the process is capable of producing within acceptable standards, you may well be flogging a dead horse in trying to implement statistical methods to control it. As we mentioned in Chapter I, if a process is incapable of producing up to standards SPC will not fix it, control it or improve it. Statistical methods will, however, tell you the process is incapable.

All too often workers are blamed for shoddy workmanship when the problem is not the workers but the process itself. You say, "What can be wrong with the process? We've got a design, people and machines. All the people need to do is run the machines to the design specifications." That may seem rather simplistic, but you might be surprised at the number of managers who think along those lines. Stop and think for a moment. What if: the machines in use were not designed for today's tighter specifications; engineering did not coordinate adequately with production in product design; workers were not trained or were inadequately trained on new machines, modified old machines or new procedures on old machines; clerical forms were not printed with adequate room for entries, leading to sloppy entries and inevitable recording and transcribing errors; and, on and on? "What if", is that for any of a vast array of common job design and development mistakes one ends up with a process that cannot produce what it is supposed to produce. If a process is incapable of meeting standards, you can hardly expect to control it within those standards.

Ascertaining if a process is capable of meeting standards is nothing more than using collected data to decide if the calculated upper and lower control limits of a process, which is stable and in control, lie within product specifications. If a process is not capable, then before SPC can be implemented, management, *MANAGEMENT*, must take action to correct the process. If management cannot or will not act to correct the process then they can expect no improvement in the quality of output from *any* quality improvement program, much less from SPC.

DETERMINING PROCESS CAPABILITY

Determining process capability is not at all hard, but it is frequently neglected. It should be done first in any quality improvement program and should be repeated on a periodic basis to ensure the process stays capable. Remember, if a process is incapable it is management's responsibility to either accept the poor quality or improve the process. The workforce cannot be held responsible for failure to meet standards if they do not have the means to reach those standards.

Let's go back to our collection of an initial set of data in the last chapter. In applying statistical methods to process control we need an initial data set to determine the $\bar{\bar{X}}$ and $\bar{R}$ for our process. From $\bar{R}$ we can estimate the process standard deviation by calculating $\hat{\sigma}$. In turn, $\hat{\sigma}$ is used to calculate the upper and lower control limits of the $\bar{X}$ chart (its factors are included in the A_2 constant for determining 3 sigma limits from $\bar{\bar{X}}$, thus $\hat{\sigma}$ may not be independently calculated on every data set).

As was mentioned in the section defining the standard deviation, $\hat{\sigma}$ is an adequate and accurate estimate of the population standard deviation, with one qualification: The process must be stable and tightly controlled. If the process is not stable and in tight control the picture of the population as shown by the $\bar{X}$ chart will not be accurate. Actual population limits will be wider than indicated by the $\bar{X}$ chart control limits.

A term often used in capability studies is "inherent process variability". Inherent process variability is equal to 6σ, which means output of a stable, in control, process can be expected to vary within a range of ± 3 standard deviations from its mean. That much variation is "inherent" in the process. We also know that for a stable normally distributed population 99.73% of the population will lie within this ± 3 sigma range from its population mean (as estimated by $\bar{\bar{X}}$). In determining if a process is capable from our initial data set the only problem becomes getting a firm fix on the standard deviation (sigma). You must decide if $\hat{\sigma}$ is an acceptable estimate of sigma.

From our initial data set we discussed recording samples in a time ordered sequence, calculating $\bar{X}$'s, R's, $\bar{\bar{X}}$ and $\bar{R}$. After these calculations data points are plotted, $\bar{\bar{X}}$ and $\bar{R}$ lines are drawn and appropriate control limit lines are determined and drawn. Once the charts are drawn the distribution of data points must be studied. If data points appear to be randomly distributed within the control limits we can reasonably assume the process is in control and thus predictable (notice we did not say capable, that has yet to be determined). If the data points exhibit wild fluctuations, trends, shifts, or points outside of limits, then there are assignable causes for process variation. Every attempt must be made to identify and eliminate the assignable causes before any further process capability study, not to mention process control, can proceed.

A few notes on capability studies. In Chapter 2 we mentioned the importance of making notes on any changes in the process during data collection. This is especially important in determining if the process is capable. In making a capability study on an operation record everything which might have a bearing on the data. You need machine data (which machine, speeds, rates), material data (where did materials come from, lot number, characteristics) and operator data (who, level of training). The study should be conducted under normal operating conditions and corrections or adjustments to machines, measuring equipment or the process itself should be avoided. Any deviation to these guidelines, and particularly failure to record deviations, can lead to confounding data.

A capability study was stalled for some time only to find that the machine operator, in his zeal to do well, was overadjusting his machine to account for minor variations in output; and no one bothered to record it. Once the problem was found and the operator was retrained on the machine operation the gyrations in the output stopped. As this aptly illustrates, you must attempt to maintain all factors affecting the process as stable and as normal as possible during the course of the capability study.

After the assignable causes have been eliminated from the process in so far as is practical and you have what appears to be a usable set of $\bar{X}$ and R charts, it is time to decide if $\hat{\sigma}$ as determined using $\bar{R}/d_2$ is an adequate estimate of sigma. If your R data points appear to be randomly distributed about and for the most part lying close to $\bar{R}$, and the $\bar{X}$ points appear to be likewise distributed randomly about and lying close to $\bar{\bar{X}}$, then most likely the process is stable and in tight control. In this case $\hat{\sigma}$ can be expected to be a good and usable estimate of sigma. However, in the more likely event that your process now appears stable and thus predictably lying within its ± 3 sigma control limits, but with data points not exhibiting the tight control within those limits, you must decide how to handle the fact that the $\hat{\sigma}$ obtained from $\bar{R}/d_2$ may well not be a good estimate of the population standard deviation. Don't panic. There are two ways to approach the problem, one requires some rather lengthy but simple arithmetic, the other simply adds in a fudge factor.

Deere and Company[*] recommends using the long hand equation for deriving $\hat{\sigma}$:

$$\hat{\sigma} = \sqrt{\frac{n\Sigma x_i^2 - (\Sigma x_i)^2}{n(n-1)}}$$

where: $\hat{\sigma} = $ the estimated standard deviation

$n = $ the total number of readings taken (not the number of samples, but all of the readings of all of the samples)

$\Sigma = $ this is a mathematical symbol which means to sum (add together) all of the following

$x_i^2 = $ the square of every reading taken (multiply the reading value times itself) for each x value from $i = 1$ (the first value) to $i = n$ (the last value)

$(\Sigma x_i)^2 = $ the sum of all of the reading values added together and then squared

The steps of the equation are: 1) take each reading from every sample and multiply it times itself (square it), add all of these squared numbers together and multiply the obtained sum by the total number of readings taken (n); 2) add all of the readings from all of the samples together and square the sum (i.e., you don't square the individual readings in this step, you square the total of adding all of the readings together); 3) subtract the value obtained in step 2 from the value obtained in step 1; 4) determine the denominator by taking the total number of observations taken (n) times the total number of observations minus one $(n-1)$; 5) divide the value obtained in step 3 by the value obtained in step 4; and 6) take the square root of the value from step 5. This may seem a rather long and tedious process, and it is, but most of today's pocket calculators with 2 memories will make short work of it. The equation does yield a very accurate $\hat{\sigma}$ for estimating the true sigma of the process.

On the other hand, according to Seigel[†], it is common practice in the automotive industry for short term process capability studies to simply use $\pm 4\hat{\sigma}$ as a conservative estimate of the parent population. This is justified statistically by noting that for a stable population the distribution of the parent population may be wider than the distribution of the averages ($\bar{X}$'s) by a factor equal to the square

[*]*Capability Study Guidelines* (Deere and Company, Reliability Department, August 1981), p. 12

[†]James C. Seigel, *Managing With Statistical Methods* (SAE Technical Paper Series #820520, 1982)

root of the sample size. So, when using small sample sizes of 4 or 5, simply adding a correction factor of $+1\hat{\sigma}$ to the upper control limit and $-1\hat{\sigma}$ to the lower control limit on the $\bar{X}$ chart will yield justifiable limits of what the process is capable of.

The goal of this whole exercise for finding a good $\hat{\sigma}$ is to predict what the limits are for the parent process population. If the $\hat{\sigma}$ determined in the initial $\bar{X}$ and R charting process looks good, use it. $\bar{\bar{X}} \pm 3\hat{\sigma}$ will yield a reasonable estimate of the limits of the parent population. If there is some question as to the accuracy of $\hat{\sigma}$, play it safe and use the Deere and Company method for computing a more accurate $\hat{\sigma}$ before calculating the $\pm 3\hat{\sigma}$ estimate of the parent population's limits. If sample sizes are kept small (4 or 5) the method outlined by Seigel of using the original $\hat{\sigma}$ but estimating the limits of the parent population at $\pm 4\hat{\sigma}$ from $\bar{\bar{X}}$ is computationally simpler with just as good results.

With the limits of the parent process output population now adequately estimated, the time required to determine process capability can be measured in microseconds. Simply look at what the process is supposed to yield (part tolerance limits, output specifications, whatever the process standards are) and look at the limits of the parent population. If the parent population's limits do not lie completely within the standards set for the process, the process is incapable of producing to standards. The monkey shifts from production's back to management's back to fix the process through whatever is necessary to make the process capable. Only then can you have process control and quality improvement. Once the parent process output population's limits are within standards, then statistical methods can be applied to the process.

You should note that there is no requirement for the parent population to be centered within process standard limits. Just as long as the parent population's limits are within standards the process is capable. It may even be advantageous under certain circumstances to have the process output centered toward one end or the other of the product specification limits. But, be wary about a population too close to limits. Feigenbaum recommends using a minimum of 50 readings in a capability study (a completely filled out Ford Motor Company chart would have 150 readings, 5 readings per sample times 30 samples). He then goes on to say that the process limits as determined from these readings should not exceed 75% of total tolerance.* This is the numerical equivalent of the method discussed by Seigel in going from a $6\hat{\sigma}$ width to an $8\hat{\sigma}$ width ($\%8 = .75$). However, if you used $3\hat{\sigma}$ limits when you mean $4\hat{\sigma}$ limits and the lower (or upper) limit of the process butts up against the lower (upper) tolerance limit, you actually have 12.5% of what should be the width of the parent population's ± 3 sigma limits hanging out in the breeze, which equates to roughly between 1 and 2 percent of your parent population outside of product tolerance limits. In other words, the process is really incapable. Feigenbaum's recommended procedure works, just remember to interpret where the population lies between the tolerance limits.

If you have any problem visualizing whether process limits are within tolerance

---

*A. V. Feigenbaum, *Total Quality Control*, 3d ed (New York: McGraw-Hill, 1983) pp. 782-790

limits, draw it out on a piece of paper or a graph. As a matter of course, that is the easiest way to communicate the results of a capability study to management. Use a line chart and mark the process tolerance limits on it and then mark the process output limits on it. The facts will speak for themselves. If the process is capable, move on to statistical control of the process. If the process is incapable, the extent of the problem is clearly visible.

NARROW LIMIT GAGING

Narrow limit gaging is a means of achieving statistical process control over typical manufacturing processes with a minimum of operator training.

From the capability study, if the process is capable we found the distribution of process output is within output specification limits. Assuming the capability study showed the process to be capable, production workers cannot start from this point and be expected to record and manipulate data on $\bar{X}$ and R charts without some appreciable training. Where go/no-go gaging has traditionally been used or can be implemented to measure product acceptability, narrow limit gaging can be used to achieve process control roughly equivalent to the use of $\bar{X}$ and R charts. Where as the use of $\bar{X}$ and R charts requires some formal training effort, narrow limit gaging can be implemented with little more than minor instructions required for operator training.

For those unfamiliar with the go/no-go gage, it may be a fixture, jig, dial indicator or some similar device to which sample items are subjected for testing. The testing procedure can be destructive or non-destructive. If items pass the test, output is accepted; if items fail the test, output is rejected: a relatively easy procedure for even unskilled workers to use.

The problem with go/no-go gaging is that, if only samples are gaged and if the gages are set to only check for meeting tolerance limits, you can run into significant errors in predicting what the process is doing. The process can shift one way or the other to where one tail of the distribution now lies outside of specification limits, yet small samples taken to monitor the process might not signal this shift for some time. Significant numbers of rejects could already be on their way to customers. The way to beat the problem while using the same go/no-go methodology is to narrow the gages. The statistical justification for this is easy to visualize. As the distance from the population mean increases (moves toward the tolerance limit) the probability of an occurance at that distance decreases. Thus, we are more likely to catch a population shift if we move the tolerance limit in for gaging purposes.

In narrow limit gaging you still need a capability study. The upper and lower specification limits (tolerance limits) of acceptable output must be greater than the 6 sigma population range (i.e., the process must be capable and stable). The gage limits are then set in from the specification limits by some factor of sigma. As a rule the tolerance is reduced by 1 to 1.2 sigma. Quite often only one tolerance limit needs to be shifted in toward the mean, as the output population actually lies toward one end or the other of tolerance limits. The following diagram illustrates what is done.

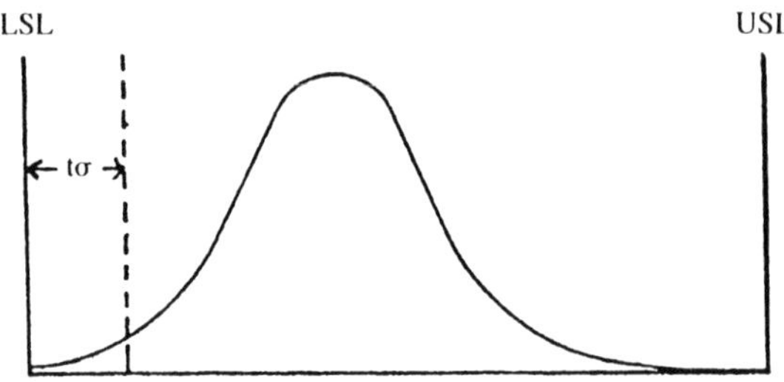

where: LSL = lower specification limit

USL = upper specification limit

tσ = amount tolerance is reduced for the narrow limit gage

Once the gages are set a plan must be in place to decide what sample sizes to use and what to do with the results of the narrow limit gaging process. Sample sizes are the same as for $\overline{X}$ and R charts, usually 4 or 5. Samples of 9 or 10 are likely to be too sensitive to process variation and can lead to many "wild goose" chases looking for insignificant problems. Remember, some shifting in the process is normal and narrow limit gaging accepts that, that is the reason for specific tolerance limits. What we are searching for are adverse shifts that show the process moving out of acceptable tolerances.

Samples are taken at regular intervals, just like in $\overline{X}$ and R charting. If more information is needed, collect samples more frequently rather than increasing the sample size. Realizing that an item rejected by the narrow limit gage may well meet product specifications, items rejected by the narrow limit gage can be re-checked for ultimate acceptance. However, the number of items rejected per sample must be recorded.

Decision criteria must be established concerning the number of acceptable rejects per sample without signaling the need for corrective action. For samples of 4 or 5 the number is usually no more than 1 reject per sample (do not use 0, you must expect some good items to lie outside the narrow limit gage specifications). If 2 items in 5 are rejected by a narrow limit gage set in 1 sigma from the product specification, the process has probably shifted and requires adjustment.

With narrow limit gaging the responsibility for the capability study, determining where to set the gage and determining decision criteria falls on supervision and/or management. All the workers have to do is know when to sample, how many items to take per sample, how to use the gage and what the decision criteria are. If a sample is within the decision criteria the worker does nothing. If the sample exceeds decision criteria the process is to be adjusted.

Supervision and management monitor the recording of samples and rejects looking for trends of frequent adjustment. Too frequent requirements for adjustments may signal the need for a new capability study and a search for problems affecting stability.

If increased accuracy in detecting process shifts is desired, sequential narrow limit gaging plans can be used with no additional effort. A sequential plan still uses the first plan, but also looks at the running results of a series of samples, subjecting the series of samples to somewhat tighter criteria. For example you could use a sample size of 5 with an acceptable gage failure limit of 1 item per sample. Then, add to this initial criterion that for the last 3 samples (totaling 15 items) there can be no more than 2 gage failures. The addition of the sequential plan to the basic plan will yield a more sensitive touch to the monitoring plan at no additional expense in time or effort.

Ott[*] recommends using any of the following 3 plans for narrow limit gaging, noting that their sensitivity is roughly equivalent to an $\overline{X}$, R chart using a sample size of 5.

Plan A	**Plan F**	**Plan D**
$n = 5$	$n = 10$	$n = 4$
$t = 1.0$	$t = 1.2$	$t = 1.2$
$c = 1$	$c = 2$	$c = 1$

where: n = sample size

    t = the number of sigmas the gage is set in from the limit

    c = acceptable number of rejects per sample

SUMMARY

This chapter looked primarily at process capability studies. It cannot be over-emphasized that before an ongoing SPC program can be implemented a process capability study must first be conducted. If a process is incapable of consistently producing output within established standards, neither the workers nor the process can be expected to show quality improvement until such time as the process is made capable through management intervention.

As a logical extension of the process capability study effort this chapter also looked at narrow limit gaging. Under situations where go/no-go gaging is applicable, narrow limit gaging programs can yield the benefits of more complex SPC programs with minimal operator training.

Chapter 4 will look at $\overline{X}$ and R chart interpretation and at a variety of chart variations applicable to certain types of scenarios.

---

[*]Ellis R. Ott, *Process Quality Control* (New York: McGraw-Hill Book Co., 1975) p. 170

IV. $\overline{X}$ AND R CHARTS

Chapter 2 showed statistical concepts and defined how to collect and chart an initial set of data on a process. Chapter 3 went on to look at that data to see if the process was in control relative to chart upper and lower control limits and capable relative to product specifications. So far we have spent little time actually looking at charts to visualize what might be seen in practice. Following are six representative combinations of $\overline{X}$ and R charts labeled, appropriately, from 1 to 6. Each chart is discussed as it is presented.

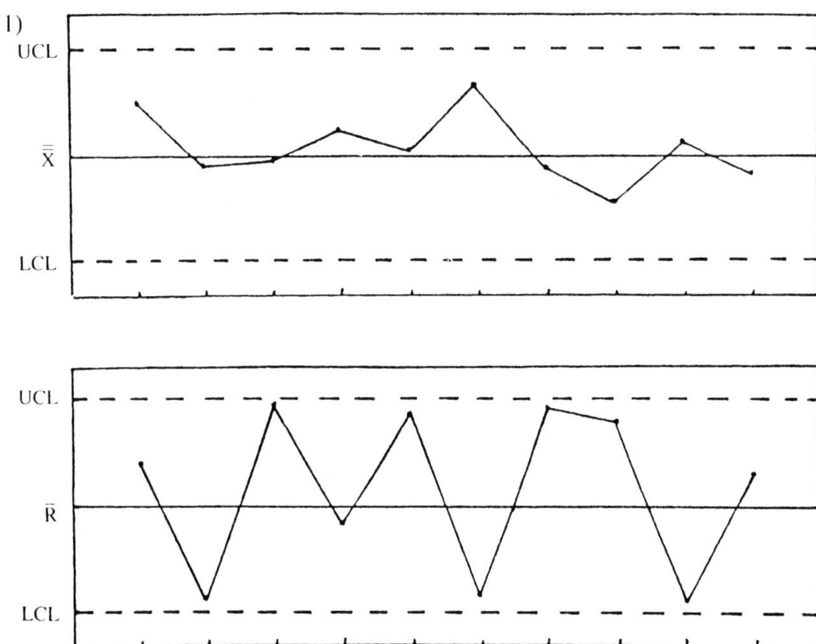

Here the $\overline{X}$ chart looks good. The individual points appear randomly located about the mean, yet most are reasonably close in to the mean. The R chart, on the other hand, appears out of control. Though no points exceed the control limits the chart still shows rather wild variation between parts in some samples. Charted points do not necessarily have to exceed control limits to indicate an out of control condition. The almost cyclical nature of the points in the R chart may indicate a specific type of problem which will be looked at in example 2.

You might ask how the $\overline{X}$ chart can look so good and the R chart can look so bad. Stop and think: the average of 4 and 6 is 5, so is the average of 0 and 10. Also remember, the LCL for R charts on small sample sizes (less than 7) is 0. So there would be only one real limit that could be exceeded on an R chart for small samples, the upper one.

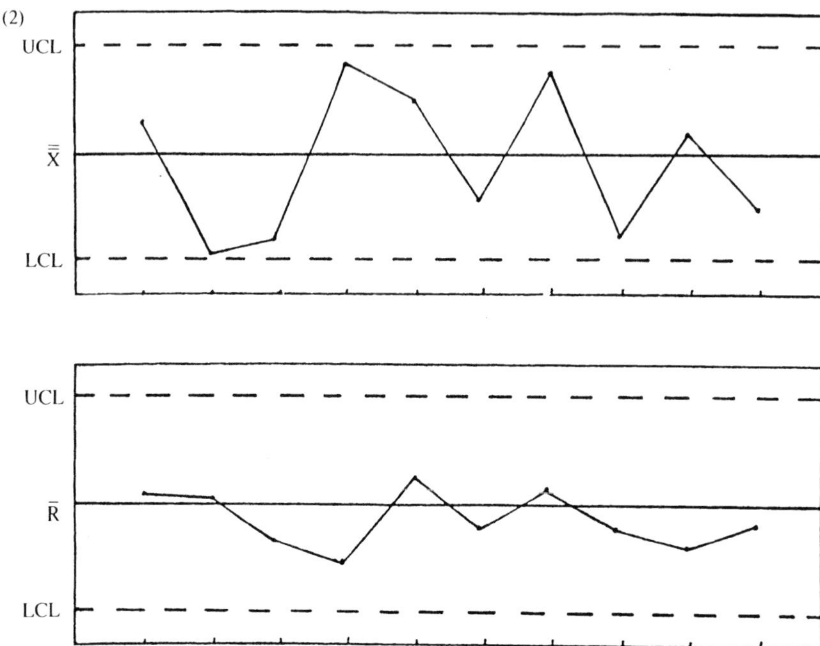

Here the R chart looks reasonably stable, no real trends, good distribution of points. But the $\bar{X}$ chart fluctuates wildly. As in the R chart from (1), no points are outside the control limits, yet less than half of the points are anywhere close to the mean. Something is obviously wrong. The distribution of $\bar{X}$ points in this chart may indicate something even more than first glance appreciation that the $\bar{X}$ chart is out of control: you may have two different processes going where you thought there was only one. Looking at either the five points above the mean or the four points below, if you considered each of these sets of points individually and recalculated new means for each set, you might come up with two separate in control sets of points. Go to the comments blocks on the charts to look for clues. A typical reason for such a distribution is two different operators on the same machine. One operator sets the machine at one setting to run, the other operator at another setting. Both outputs are normally distributed, yet neither distribution is where it should be: centered on the process mean. A little retraining on machine operation would not only center the distribution, but even tighten the output population's distribution.

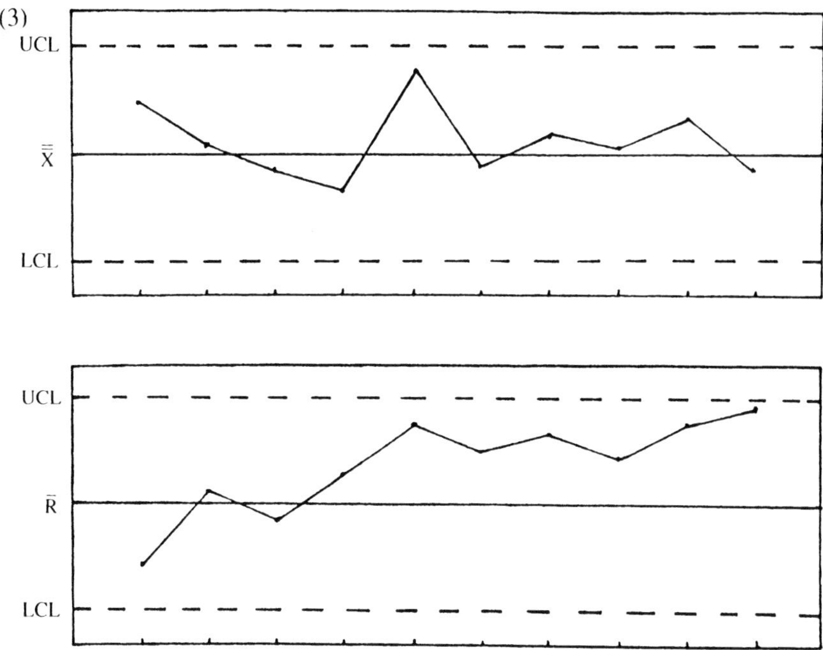

$\overline{X}$ looks good. Point 5 appears somewhat out of context, but not abnormal. It is within limits and the rest of the points are fine. The R chart, however, shows a trend upward and you do not need to watch it punch out the top of the upper control limit before taking action. Most likely something in the process is wearing out or becoming fatigued.

Looking at control charts, this is a good place to insert a brief discussion of runs. Runs are special categories of trends. Several points in a row of increasing value or decreasing value are a run. They may or may not cross the center line of the chart. Several points in a row on one side or the other of the centerline (either the $\overline{\overline{X}}$ or $\overline{R}$ line, depending on the chart) may also be considered a run, regardless of consistent increasing or decreasing nature. Runs indicate either the $\overline{\overline{X}}$ or $\overline{R}$ is shifting or the whole process is shifting to a different level. Usually a run of seven points is considered critical in SPC applications. A run of 7 points indicates a definite shift in the process with less than 1% probability of error for normal control charts (30 to 40 plotted points). For our example charts here, we will not illustrate runs due to space limitations. None-the-less, for charts in use, evaluation of runs should be normal procedure as they are a powerful tool in maintaining control over the process.

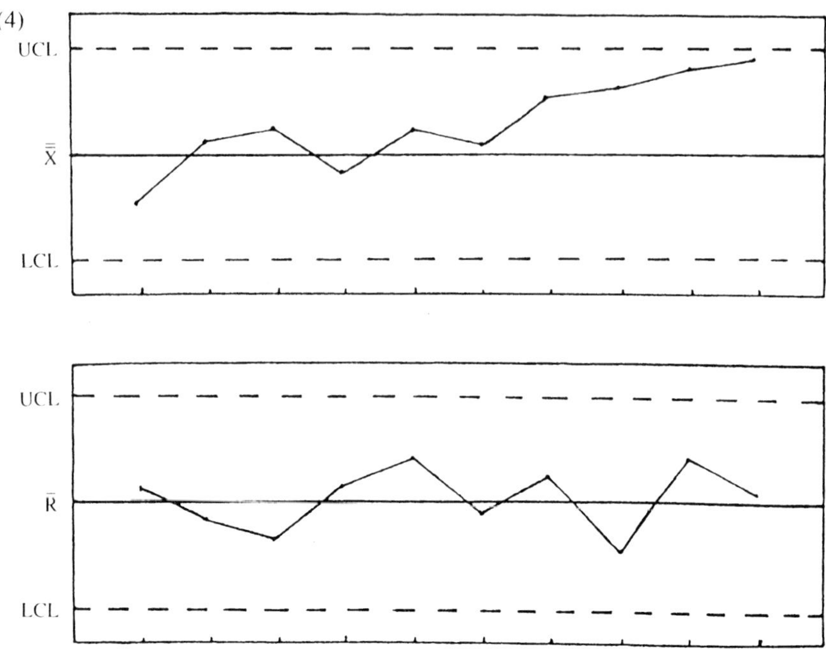

The R chart for (4) looks stable: no trends, pretty even distribution of points. The $\overline{X}$ chart, though, appears headed out the top of the chart. In a typical manufacturing operation this type of trend is characteristic of a machine adjustment slowly slipping and it is time to readjust it.

The example charts here show upward trends as illustrations. Down trends occur with equal frequency, depending on the type of process involved. In addition, trends may not be all bad. In the case of an R chart a downward trend can indicate a more uniform process. For an $\overline{X}$ chart a trend can show movement toward a more optimal level of output. In the event of good trends, once the trend has stabilized a new capability study would be conducted to reset chart control limits at the new output levels.

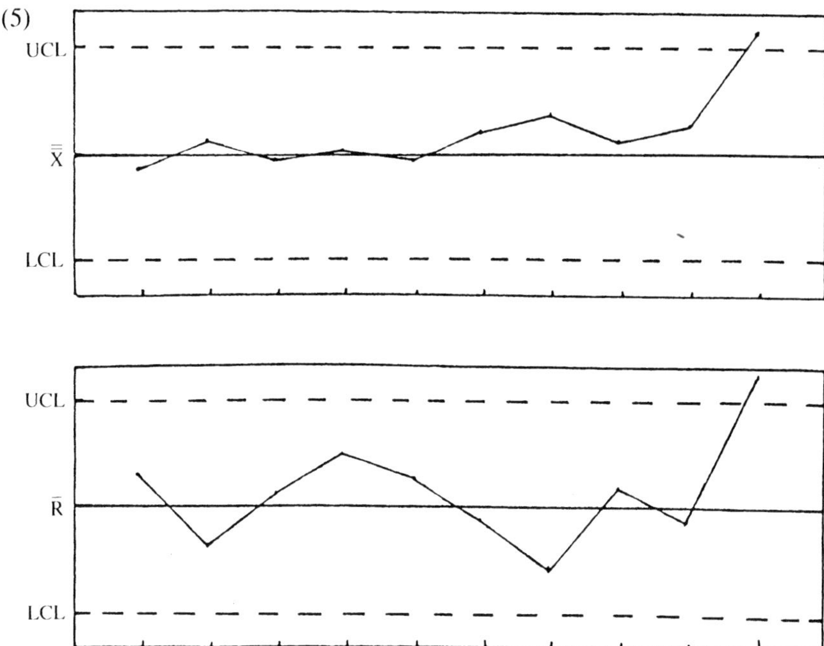

These two charts are easy. Both have points exceeding control limits. In practice it may be one or the other chart which goes out of control and not both. If both do go out of control they may not both go at the same time like the charts in the illustration. An out of control situation need not be preceded by any particular pattern of points; sometimes when a chart goes out of control it just goes. The important thing to remember is that when either chart does have a point exceeding control limits it is time to take action and find the cause of the outage and fix it.

Don't go off the deep end in charging out to shut down production over a point out of limits. There may be some logical explanation for the outage. Start up problems, tool breakage, or some similar event could have unpreventably caused a momentary process abnormality. Check the comments blocks on the charts first. Barring any of these brief knee-jerk type of symptoms, points out of control mean problems. Rejects will shortly be coming off the line and the problem is not going to go away. Action must be taken to identify the cause(s) of the problem and rectify it (them).

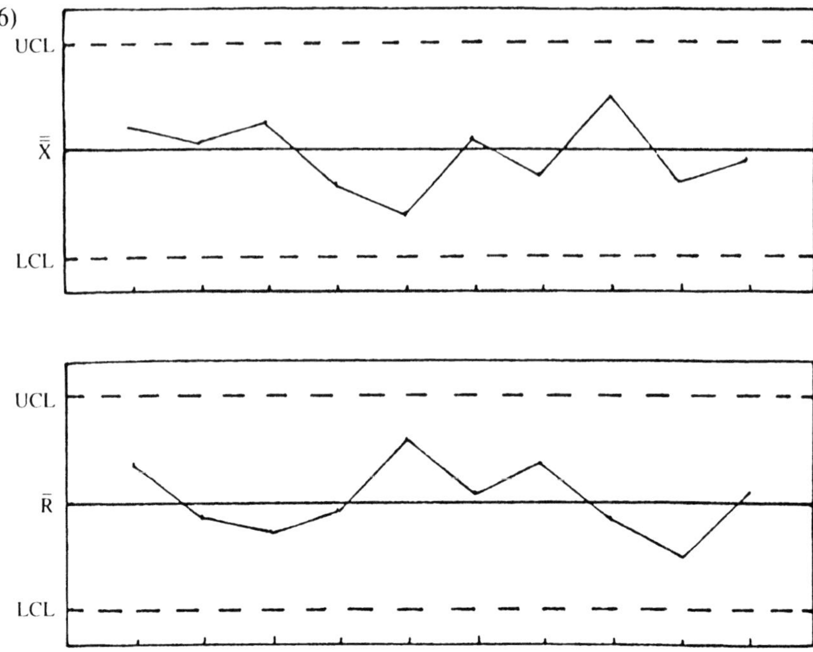

Consistent charts like these would make for nice quiet days at work. All of the points are within limits, no trends, normal peaks and valleys, and a good random distribution of points. A nice stable in control process.

You *can* get processes to this point. They may not stay this nice all of the time. That is the reason for the emphasis on SPC as a continuing and on going program. New problems will arise and with SPC you become aware of them when they happen. Quicker problem identification means quicker problem fixes, less waste and higher quality.

You have seen six representative pairs of $\bar{X}$ and R charts. Charts are simply a means of testing your process. In scientific or statistical terminology you establish a hypothesis, a belief, that the process is capable of producing what it is suppose to produce, consistently. The charts are a method for proving or disproving that hypothesis on an ongoing basis. If your charts are stable with randomly distributed points lying within chart limits, with a high degree of certainty you can accept your hypothesis. If the charts show trends, runs or outages, your hypothesis probably has holes in it that need to be investigated and fixed.

VARIABLES VS ATTRIBUTES

There are two ways to analyze the quality of output. One measures physical characteristics such as diameter, circumference, depth, height, width and so on. These measurable physical characteristics are called variables. The second way to measure quality is through attributes. Where variables have a range of values, attributes have only two: yes or no.

Variables fit the classic discussion of applying statistical methods to traditional manufacturing processes. Output items are accepted or rejected based on physical dimensions because output items are parts of a larger entity, and the parts must all fit together. The tolerance limits of output are based on its critical limits as an input item in the next larger assembly. These types of items fit the familiar techniques of statistical process control: small sample sizes and the use of $\overline{X}$ and R charts. Up until this point in our discussion about statistical methods we have implicity been discussing SPC as it applies to variables.

With some minor conceptual variations and chart adaptations statistical methods application to process control can be expanded from the narrow realm of controlling variables to the virtually all encompassing realm of controlling attributes. For all of those processes where quality is not measured in physical dimensions there are ways to measure quality as a function of attributes.

Some terms associated with attributes are: pass/fail, go/no go, conform/nonconform and discrepancy/no discrepancy. Attributes have only two values, yes or no, they either pass or don't pass. There are no degrees of passing and this is frequently one of the hardest parts of trying to control attributes. The determination of passing or failing is often the subjective judgement of an inspector. It is extremely important to define as explicitly as possible what constitutes accept/reject standards to remove, in so far as is practical, as much of the subjective element of the decision as possible.

Attributes can be applied to all kinds of processes from visual inspection of semi-finished or finished goods to transactions at a bank teller window. In looking at goods coming off a manufacturing line typical attribute items might be the presence of dings, chips or runs in the paint, lights that don't work, electrical circuits failing continuity tests and labels improperly positioned. Attributes can be used to measure customer satisfaction by noting the number of customer complaints per number of items sold or number of customer transactions conducted. The different types of attribute applications are endless.

One good thing about attributes is that the information needed to track them is probably already available and simply needs to be reformatted. Quality Control functions likely already collect inspection data on semi-finished or finished goods. Customer complaints data can be correlated to the number of units sold or serviced or customer transactions conducted. Where needed attribute data is not available it is usually relatively easy and cheap to obtain as compared to variable type data.

There are four primary ways to chart attribute data. First, for samples of not necessarily equal size the p chart is used to track the proportion of units nonconforming. Second, for samples of constant size the np chart is used to indicate the number of units nonconforming per sample. Next, for samples of constant size the c chart can be used to follow the number of defects per lot. Finally, the u chart is used to note the number of defects per unit from samples not necessarily of constant size. The next section will introduce and briefly cover the application of each of these charts.

OTHER TYPES OF CHARTS

s Chart. Before moving into charts for use with attribute data there is one last chart to mention in conjunction with variables. s charts are used in place of R charts for measuring process variability in some instances, usually when using larger sample sizes. s stands for sample standard deviation and, as you might suspect, it is mathematically more complex to calculate than the sample range (R). The formula for calculating s is:

$$s = \sqrt{\frac{\Sigma x_i^2 - n\bar{X}^2}{n - 1}}$$

where: x_i = each sample measurement from i = 1 (the first value) to i = n (the last value)

$\bar{X}$ = the sample mean

n = the number of items in the sample

The steps for calculating s are: 1) square each sample measurement and sum (add) all of the squared values together; 2) find the mean of the sample, square it, and multiply the squared value by the number of items in the sample (n); 3) subtract the value obtained in 2 from the value obtained in 1; 4) divide the value obtained in 3 by n minus 1, and; 5) take the square root of the result of step 4.

While s is a somewhat more efficient measure of process variability it is more complicated to calculate and it will mask single items in a group that are way off the mark in relation to the other items in the sample. This last fact alone bodes caution in deciding when to use an s chart. s charts would be used where one would expect more gradual as opposed to sudden shifts in variability. The efficiency gained from the more complex calculations would likely be worthwhile only for less frequent but larger samples (samples of size 10 or more).

The method of charting the $\bar{X}$ and s chart is the same as for $\bar{X}$ and R charts, except that we change all of the constants. The new control limits for the X chart are:

$$UCL = \bar{\bar{X}} + A_3\bar{s}$$
$$LCL = \bar{\bar{X}} - A_3\bar{s}$$

where $\bar{s}$ is the average of the s's for all the samples. The control limits for the s chart are found using the equations:

$$UCL = B_4\bar{s}$$
$$UCL = B_3\bar{s}$$

Drawing the $\bar{X}$ chart is the same as before. Drawing the s chart is also the same as drawing the R chart except now you use s in place of R. Estimating the process standard deviation from $\bar{s}$ is through:

$$\hat{\sigma} = \bar{s}/c_4$$

You can even use the same $\bar{X}$ and R chart form that you have set up for other applications if you desire. You'll need to mark over the R chart annotations and may need to record your sample data on another form due to the larger sample sizes. Other than the more complicated calculations the use of the s chart is the same as the use of the R chart. In parallel with the $\bar{X}$ chart you use the s chart to evaluate the process for initial control, then check the process for capability. Once control and capability are established develop a program for using the charts for long term process control.

p Chart. The p chart is the first attribute type chart we will look at. p stands for the proportion of nonconforming items in a sample of items.

When looking at attributes, which this and the next 3 charts will be concerned with, each item of each sample is inspected for defects. The defects can be functional defects (such as failing a continuity test) or simply appearance defects (such as labels being improperly positioned on a bottle). The inspection can be for a single type of defect or numerous types of defects. However, for the p chart whether one or many defects are found the item is counted as nonconforming (defective) only once. For example, if an order form has 5 entry errors it is one defective order form, not 5.

Sample sizes for attribute charts are larger than for the $\bar{X}$ and R charts used with variables. There are formulas for calculating minimum sample sizes when certain confidence intervals are demanded on how accurate your final answer needs to be. The problem with these formulas is they often demand unrealistic sample sizes and thus do not fit the real world needs of process control work. To resolve this conflict between theory and application, according to Ott, when a single shift's daily output is less than 500 it is best to go with a 100% inspection, otherwise use initial sample sizes of no more than 150. This sample size can then later be raised or lowered based on experience.* Ford Motor Company, on the other hand, recommends using samples of 50 to 200 or more. †The key to sample size is that the sample should be large enough to have several nonconforming items per sample. The average number of nonconforming items per sample should be greater than 5 or 6.

Determining how often to sample is similar to that for $\bar{X}$ and R charting. Sampling should be often enough to detect changes in process output and over short enough time periods to provide meaningful feedback. The length of the time period may conflict with the need for large sample sizes, so there may need to be some trade-off between sample size and length of time required to collect the sample. A lower limit on sample size is 50. Samples smaller than 50 will lead to peculiar and unreliable results due to the fact that you are dealing with fractions and percentages. At the same time, sample sizes should not be so large as to cover too long a period of time, which can either mask process changes or provide less than timely notice of process changes.

Just a couple of more comments on the sampling process itself. Different inspectors can cause wide variations in the number of defectives found in a lot. As was explained to before, much attribute data is subjective in nature: it is the opinion of the inspector. It is extremely important to define as specifically as possible

---

Process Quality Control, pp. 78-79

†*Continuing Process Control,* p. 36

what qualifies as a defect to try to minimize this subjective judgement factor. Next, a number of characteristics may vary together: if one attribute is defective another related attribute may be defective also. This is known as a correlation between the characteristics. In an effort to keep the inspection process uncomplicated, for strongly correlated attributes it may be adequate to inspect for only one of them. Finally, sample sizes for p charts do not have to be of the same size, but you should try to keep them within $\pm 25\%$ of the mean sample size. If you do not it will not change the calculations; however, when we get into the charting process it may add to the calculations.

The p data collection and charting process is very similar to $\bar{X}$ and R charting, except you only have one chart. The next page shows a representative p chart form taken from a Ford Motor Company training manual. This particular chart can be used for p, np, c or u charting by checking the appropriate symbols used or marking through the ones not used. For each point you need 2 inputs. First you need the size of the sample (n) and then you need the number of nonconforming items in the sample (Ford uses the symbol "np", we will use the symbol "d", meaning defective, to avoid confusion with other symbols). The proportion defective is then computed by dividing the number of defective items (d) by the sample size (n).

While collecting and recording data for the attribute charts, just as with the $\bar{X}$ and R charts, it is important to keep a process log noting anything which could cause an aberration in the process. These comments will later be invaluable in interpreting the data points and charts as a whole.

As with our previous charting discussion, to start a charting program you need an initial data set consisting of 25 to 30 samples over a representative length of time (i.e., all shifts, several samples per shift, if appropriate, and over at least several days). Each item in each sample is inspected, noting the total number of items inspected and the number of items inspected which were found to have defects.

From this initial data collection the calculation and plotting of data points for the p chart is just a variation of the previous charting methodology already covered. Having recorded the sample size and number of defectives and then calculated the proportion defective for the individual samples, choose an appropriate scale for the chart where each of the calculated p's will easily fall within the chart scale, then plot the points. Next you need to calculate the average (mean) value for p, $\bar{p}$. $\bar{p}$ is found by adding together all of the p values and dividing by the number of samples taken.

Now you are ready to find the chart control limits. Just as before, $3\hat{\sigma}$ control limits are used. The formula for this $\hat{\sigma}$ is:

$$\hat{\sigma} = \sqrt{\frac{\bar{p}(1 - \bar{p})}{\bar{n}}}$$

where: $\bar{p}$ = the average proportion defective for all of the samples taken

$\bar{n}$ = average sample size found by summing all of the sample sizes and dividing the sum by the number of samples taken

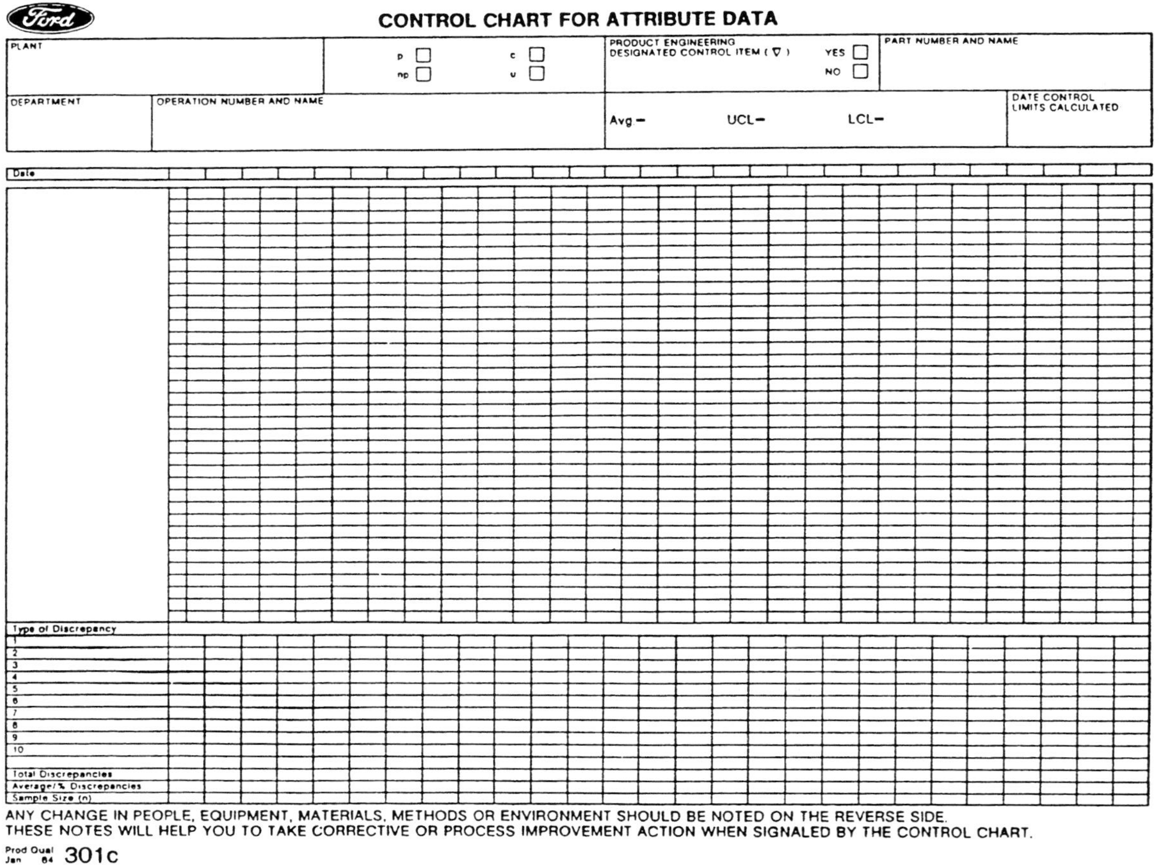

Reprinted with permission from *Continuing Process Control and Process Capability Improvement. A Guide to the Use of Control Charts for Improving Quality and Productivity for Company, Supplier and Dealer Activities.*
Ford Motor Company Booklet No. 80-01-251, July 1983

ATTRIBUTE CONTROL CHART FORMULAS

Nonconforming Units	Nonconformities

Number

(Subgroup sizes must be equal.)

np Chart

$$UCL\ np,\ LCL\ np = n\bar{p} \pm 3\sqrt{n\bar{p}\left(1 - \frac{n\bar{p}}{n}\right)}$$

c Chart

$$UCL\ c,\ LCL\ c = \bar{c} \pm 3\sqrt{\bar{c}}$$

Proportion

(Subgroup sizes need not be equal.)

p Chart

$$UCL\ p,\ LCL\ p = \bar{p} \pm 3\sqrt{\frac{\bar{p}(1-\bar{p})}{\bar{n}}}$$

u Chart

$$UCL\ u,\ LCL\ u = \bar{u} \pm 3\sqrt{\frac{\bar{u}}{\bar{n}}}$$

DATE	TIME	COMMENTS

ACTION
On Special Causes

- Any Point Outside of the Control Limits

- A Run of 7 Points — All Above or All Below the Central Line

- A Run of 7 Intervals Up or Down.

- Any Other Obviously Non-Random Pattern

ACTION INSTRUCTIONS

1.

2.

3.

4.

5.

Prod Qual
May 83 301h (reverse)

Reprinted with permission from *Continuing Process Control and Process Capability Improvement.*
A Guide to the Use of Control Charts for Improving Quality and Productivity for Company, Supplier and Dealer Activities.
Ford Motor Company Booklet No. 80-01-251, July 1983

It is a pretty simple equation. First subtract $\bar{p}$ from 1 and then multiply what is left by $\bar{p}$. The result of this is divided by the average sample size ($\bar{n}$) and then take the square root of the final value. With $\hat{\sigma}$ now calculated the chart control limits are:

$$UCL = \bar{p} + 3\hat{\sigma}$$
$$LCL = \bar{p} - 3\hat{\sigma}$$

There are no constants to bother with. Simple and straight forward. Draw the $\bar{p}$ line on your chart and highlight the lines for the upper and lower control limits.

One note here concerning the lower control limit. You cannot have less than zero defectives. However, using any of the lower control limit (LCL) formulas for attributes you can calculate a negative limit. If that happens simply consider the lower limit as being zero.

Your initial chart now plotted, evaluate the position of the plotted points. What you actually have here is a binomial distribution, so called because you can only have 2 values for each item inspected, in this case pass or fail. When $\bar{p}$ is greater than 5 or 6 the binomial distribution closely approximates the normal curve. With this in mind, and remembering back to our discussion on the distribution of points in a normally distributed population, about 2/3's of the plotted points should lie within $\pm 1\hat{\sigma}$ from the mean (the middle 1/3 of the chart). Only about 5% of the points should be close to the control limits.

The same criteria is used for evaluating the points on a p chart as for the $\bar{X}$ and R charts. There are a couple of items though that require a little closer scrutiny. If the points or limits look out of place, first check the calculations for errors and then look for errors in plotting. If the points seem to be lying closely bunched about $\bar{p}$ or there are too few points within the middle 1/3 of the chart, you may be looking at 2 different process inputs rather than 1. One additional reason for too many points lying close in to $\bar{p}$ is that the data may have been edited, with unfavorable data either left out or altered.

Having checked for calculation and plotting errors, points near or exceeding the control limits require some extra attention. Look at the sample size associated with each of these points. If the sample size exceeds or is less than the average sample size ($\bar{n}$) by more than 25% you must recalculate the control limits for each of these samples individually. To do this substitute the actual sample size (n) into the equation for $\hat{\sigma}$ in place of the average sample size ($\bar{n}$). $\bar{p}$ remains the same. Then calculate the new $3\hat{\sigma}$ limits for that point. If the point now lies within limits, it is in control and evaluate its position along with the rest of the in control points. If the point is still outside of limits it is out of control and should be investigated. Other points near the control limits (sample sizes within $\pm 25\%$ of $\bar{n}$) are considered properly plotted.

With points near the control limits now verified and limits adjusted where necessary, there are two remaining reasons for points being out of control. First, the process got better (lower limit exceeded) or worse (upper limit exceeded) in relation to that point. In either event the process needs to be evaluated for possible improvements. Secondly, the measurement system may have changed, to include people, tools and tool calibration. You would want to check this second reason first. Whereas in the $\bar{X}$ and R charting the emphasis in point position evaluation is on production process shifts, attribute charting also places increased emphasis on shifts in the inspection process.

Trend and run analysis for p charts is the same as for $\bar{X}$ and R charts except when the expected value of the number of defectives ($\bar{n}$ times $\bar{p}$) is small. Where $\overline{np}$ is less than 5 a run of 8 or more points may be needed to signal a down trend. Other than this exception, a run of 7 points is still considered critical for signaling a shift.

After having evaluated your chart points and having taken action to find and eliminate assignable causes of variation associated with those points, process capability needs to be ascertained. There are two ways to set process control limits for evaluating process capability. First, from your initial data set you can remove all out of control points for which action has been taken to remove the cause of the out of control condition. The control limits are then recalculated based on the remaining data points. The second way to calculate new process control limits is to simply gather a new set of data (25 to 30 new samples). This second method is probably more accurate and should be followed if significant changes were made to the process as a result of the first p chart.

Your chart now reflecting all points in control, the process capability is $\bar{p}$. If this proportion defective is acceptable to management the chart control limits are used to set the process control limits for continuing process control. If this $\bar{p}$ level is not acceptable to management it becomes management's responsibility to act to change and improve the process.

np chart. np charts are identical to p charts in every respect except rather than looking at the proportion of defective items in the sample you are looking at the number of defective items in the sample. Also, for p charts you could have varying sample sizes, np charts use constant sample sizes. np charts are used in preference to p charts where the average number defective in a lot of set size is more meaningful than the proportion defective.

Gathering data for np charts is the same as for p charts. Samples for the np charts are generally smaller, between 50 and 100 and, again, must all be of the same size. Samples should be large enough to find several defects per sample and frequent enough to provide meaningful feedback to the system.

The process average is found by:

$$\bar{d} = \frac{\Sigma d_i}{k}$$

Where: d_i = the number of defects found in each sample from i = 1 (first sample) to i = k (the last sample)

      k = the number of samples taken

The estimate of the process standard deviation is:

$$\hat{\sigma} = \sqrt{\bar{d}\left(1 - \frac{\bar{d}}{n}\right)}$$

where: n = the sample size

Control limits are calculated at:

$$\text{UCL} = \bar{d} + 3\hat{\sigma}$$
$$\text{LCL} = \bar{d} - 3\hat{\sigma}$$

After plotting the data points and drawing the $\bar{d}$, UCL and LCL lines, evaluate the chart for process control. The only difference here from the p chart is that there are no adjustments to the control limits due to sample size, all of the samples are the same size. Once the process is in control evaluate it for process capability. The process capability is $\bar{d}$. If $\bar{d}$ is acceptable in the eyes of management, press on with controlling the process. If the process is incapable, it is management's job to make it capable.

c Chart. The c chart measures the number of defects per lot, as opposed to the number of items defective found with the np chart. You count each defect instead of each item with defects. It is usually used with continuous flow type operations, such as cloth or sheet material manufacturing where defects are expressed as the number of defects per unit length or per unit area (e.g., per 100 feet or per square yard). The c chart is also used when inspecting individual items that potentially may have defects originating from several different sources. Examples here could be paperwork that flowed through several offices or a manufactured item that passed through several manufacturing processes.

As in samples for np charts, sample sizes for c charts must be of equal size. Choosing the size and frequency of samples is the same as for other attribute charts: the sample size should be large enough to be meaningful and to allow for several defects per sample, the samples should be frequent enough to provide timely feedback to the process.

With sample data gathered, record and plot the number of defects in each sample. The process average, $\bar{c}$, is found by adding the number of defects found in each sample together and dividing this sum by the number of samples taken. The chart control limits are found by:

$$\text{UCL} = \bar{c} + 3\sqrt{\bar{c}}$$
$$\text{LCL} = \bar{c} - 3\sqrt{\bar{c}}$$

Once the $\bar{c}$ and control limit lines are drawn evaluate the process for being in control just as with the np chart. When the chart reflects all points in control the process capability is $\bar{c}$, the average number of defects in a sample of fixed size n. The process capability is then compared against management's expectations.

u Chart. The u chart is a variation on the c chart and measures the number of defects per inspected item from samples which can be of varying size. Defects are expressed on a per unit basis in view of the possibly varying sample size.

While samples do not have to be of equal size, maintaining them within ±25% of the average sample size simplifies calculations. Record the total number of defects found and the number of units inspected per sample. An example here would be inspecting 40 boxes with 24 bottles in each box. The number of defects found would be the total number of nonconforming bottles. The number of units inspected would be 40 boxes. With the number of defects and number of units for each sample recorded, calculate u, the number of defects per unit, for each sample. This is found by dividing the number of defects in the sample by the number of units in the sample. From our example of bottles and boxes, say you found 80 defective bottles in 40 boxes. u for this sample would be 80/40 = 2 defects per box.

The process average, $\bar{u}$, is found by summing the u's for each sample and dividing by the number of samples taken. Control limits are calculated using:

$$UCL = \bar{u} + 3\sqrt{\bar{u}/\bar{n}}$$
$$LCL = \bar{u} - 3\sqrt{\bar{u}/\bar{n}}$$

where: $\bar{n}$ = the average sample size

If you have not already done so, all of the u points would be plotted at this time along with drawing in the lines for $\bar{u}$ and the control limits. As with the p chart, for those samples whose size is greater than 25% above the average sample size or less than 25% below the average sample size calculate supplemental control limits for those points. This is done by substituting the actual sample size, n, into the control limit equations in place of $\bar{n}$. From here you evaluate the process for process control and, once in control, for capability the same as with the other charts. The process capability, once the process is in control, is $\bar{u}$, the average number of defects per unit.

SUMMARY

This chapter looked first at interpreting $\bar{X}$ and R charts and then moved into the area of comparing variables and attributes. $\bar{X}$ and R charts are used to track and control variables. Attributes are tracked and controlled using an appropriate attribute chart: the p, np, c, or u chart. The attribute charts are quite similar to the $\bar{X}$ and R charts except that they are used individually rather than in pairs. You still find a process average and upper and lower control limits. Using the attribute charts to evaluate process control and process capability follows exactly the same procedure as when using the $\bar{X}$ and R charts.

Chapter 5 will tie all of these first 4 chapters together. We will look at how you integrate all of this material into a coherent package to develop an effective SPC program.

V. SPC PROGRAM DEVELOPMENT

The mechanics of implementing a statistical process control program are just like those for implementing any other program in a company. It takes planning, time, effort and money. And just like most other programs, for management to accept the idea the profitability of the program must be its major selling point. In presenting an SPC program changes in machines or product design should be downplayed, though once implemented SPC may indicate needed design or process changes. The whole thrust of SPC should emphasize using what you already have to work and manage smarter. SPC will give both workers and managers the tools to achieve higher productivity through higher quality and reduced waste.

In view of SPC program development following the same pattern as general program development, this chapter will not emphasize program design planning, per se. It will stress, rather, factors and pitfalls peculiar to SPC programs.

SELL THE PROGRAM

For any program to get off the ground, much less be successful, it must have the support of management. Gaining the acceptance of supervisors and workers comes later; management must accept the program first. In the big scheme of things, SPC really is management's program. Management must learn the statistical concepts and charting techniques just like everyone else, because it is those charts that tell them what the business is producing and what it is capable of producing. The charts can yield a major selling point of a company's product, or they can indicate process revisions needed to stay competitive.

With the impact that SPC should have on management practices it is extremely important that management understand that the type of support needed is not financial support. Though the project will need some modest financing, management often interprets requests for support as financial and/or verbal support. With an SPC program management must become intimately involved. They must learn the statistical techniques, actively support and insist on their implementation, and use the statistical results not only to sell the product but to take management action on the process when indicated. If management does not become so involved an SPC program does not stand a chance of success.

Again, the selling point of SPC to management is profitability. A functioning SPC program will increase quality and reduce waste, its gains can be measured in dollars and cents. It is a new and valuable management tool to be ingrained in company management practices. To realized SPC's highly profitable potential management must marry its practices to the SPC program.

Once upper level management has bought an SPC program everyone else that will be involved in implementing the program must be won over. Just as any program can languish from lack of management involvement, many a program has fallen by the wayside due to middle management, supervisory, or worker resistance to change. An SPC program is no different. With SPC a lot of people will initially panic at the thought of having to learn statistics. The simplicity of the statistics involved should be stressed. After all, there are only a dozen or so terms to be mastered and for any particular work station the mechanics of charting really are simple. Chart evaluation is mainly common sense.

The emphasis to everyone from mid-management down should be on working smarter and on reducing traditional tensions associated with quality problems. SPC gives mid-management and supervisors a quicker indication of existing problems before they are spotted by traditional quality control methods. The charts also provide notice of rising problems through trend and run analysis. Once an SPC program is in place and working it presents the whole mid-level management group, those people most closely associated with production, timely information to resolve quality and production problems *before* those problems show up as finished goods quality problems and waste. Not only can SPC spot problems, but through knowledgeable chart interpretation and attention to chart comments the charts can, more often than not, lead management and supervision directly to the cause. Statistical process control *is* working smarter.

Next to selling upper management on SPC, the people most hesitant about statistics will probably be those people involved in production, the people doing the work who will also have to collect the data and work up the charts. Initially they will likely take the stand that now you not only want them to do their job, you want them to be statisticians. The emphasis in gaining worker support for SPC must be on making their job easier. Every organization, to varying degrees, stresses quality workmanship. When quality problems show up in finished goods the arguments and counter-arguments concerning the blame for the problems inevitably roll down hill to production. The SPC charts give production workers the proof of the quality of their work. The charts tell when processes need to be adjusted. They can show predictability of output and they can also prove through that predictability whether the process is in fact capable of meeting management's quality demands. For the workers, SPC can help eliminate a lot of the stress involved with quality problems and provide a more harmonious atmosphere for working with management and traditional QC functions to resolve quality difficulties.

As you can see, each level in a business must be separately addressed and convinced of the benefits of statistical process control. What SPC offers each level is somewhat different. With SPC integrated through all levels the result is higher quality, reduced waste and less strife between levels. The major difference between SPC and other programs from the standpoint of initially selling the program is that you are dealing with statistics. For most people statistics is not a normal adjustment to their job, it is a totally foreign concept that may require a more than average convincing effort to gain acceptance. In addition, this is not a run-of-the-mill program that management can say, "Fine. Let's go with it and keep me posted." Management must understand they are part of the SPC program. SPC will demand their action and will influence and become a part of both their routine and non-routine decision making.

DEVELOP A PLAN

A good plan for implementing statistical methods is just like a good plan for any other project: it gives the program direction and purpose and establishes milestones. The first thing you have to do is keep your ambitions in check within achievable goals. Half baked plans are like half baked cakes, they are not very palatable to those that have to eat them.

Training is the first step in an SPC program and we will delve into that specific topic in the next section. First, though, you have to decide who to train and in what order. Then you have to come up with a training lesson plan. As far as who to train and in what order, it would be logical to start at the top of the organization and work down. Once upper level management gets educated and gets hungry for statistical information to use in decision making, their desire becomes the motivation for the next level down to get with the program. The same logic follows for each successive level down to the production floor. If you try training from the bottom up production workers will be creating charts that the higher ups do not know what to do with. Then the workers, upon whose support the program emanates, will get frustrated and sink the program. Also, be wary of getting in a hurry. If by the time you get workers trained and the program actually running there is some concern about upper levels being rusty on concepts, short refresher courses are relatively quick and easy to conduct.

Depending on the size of the organization, directing the program at successive segments of the organization is probably prudent. There are two basic reasons for this line of thinking. First, if the program is overambitious in trying to train everyone and implement SPC throughout an organization all at once, the efforts will lose focus by being spread too thin. Secondly, even if it was possible to implement SPC simultaneously throughout an organization, you can overwhelm management with a thousand ''fires'' that demand their attention. Keep the program manageable by working one area at a time. Remember, SPC is an ongoing program. It is never ''done'', so to speak. As such there is no reason to try to suddenly drown in it. A ''gradual'' implementation should be the watch word.

What area should be chosen first for SPC implementation? There are two philosophies here and either one is fine. First, you might choose an area that appears will have few problems with SPC implementation. The rationale here is that this would be the pilot program and provide an opportunity to get the bugs out of the plan with a minimum of difficulties. The other option is to use the Pareto chart method to pick an area that appears to have the most to gain from SPC, an area rife with quality problems. This is a sort of ''jump in and get your feet wet'' plan. This second idea may be more difficult to follow, but the potential benefits may well be worth it. (The Pareto chart will be explained in the section on feedback.) Of course, you can just arbitrarily pick an area to start in. In the long run which area you start in is strictly a subjective management decision. All areas can benefit from SPC implementation. Ultimately the idea is to get SPC implemented throughout the organization.

In summary, whatever direction chosen to follow in implementing a statistical methods program, some careful initial planning is a must. The areas affected and the people affected need to be brought together in the planning phase. To varying degrees, everyone involved needs to be told what is coming to smooth the way for later training and implementation. The plan outlines where you are going, how you are going to get there, and when you plan to arrive. Disseminating the plan to those people who will be swept up in the plan eliminates surprises and will allow them to ease into the transition to SPC.

EDUCATE EVERYONE INVOLVED

Coming up with an education plan is one of the more time consuming elements of not only planning the SPC program but of initially getting it started. For anyone who has ever taught a class in anything, you know that the getting ready to teach is far more time consuming than the teaching. However, for SPC you do not necessarily have to write your own lesson plans nor, if you so desire, do you have to do your own training.

There are numerous consulting firms that have developed quite comprehensive training programs for SPC. Their ads can be found in trade journals and in any of the quality control journals and magazines. Most of these ready made training programs are flexible in adapting to any specific application and are probably well worth the expense just in the savings in time, if not the headaches. Even if you want to develop your own training plan, obtaining copies of other training plans can terrifically shorten the development process.

If you decide to go with a training program developed by a consulting firm, search out a number of different companies and carefully evaluate what they have to offer. Their plans may be designed for them to do the training or may offer everything for you to do the training. Look at the texts, workbooks and training aids that come with the programs. Are they easy to understand and follow? Are their examples easily translated or related to what goes on in your line of business? Do not buy a plan because it worked for someone else, buy it because it fits the needs of the company.

Also, do not buy more than you need. Statistical data on a process can be used for many things in addition to strictly process control. It can be used to compare variability between similar machines or processing lines to determine if there are statistically significant differences between the outputs. Statistics can be used to evaluate differences in processing methods or to determine the impact that the varying of production factors will have on output. But, by and large, production workers do not need to learn these concepts or techniques, neither do most managers. These types of uses of statistical data are more in the realm of the use of statistics by traditional Quality Control functions. Keep your training program focused on the needs of the majority. If you have need of these additional uses of statistical information, gather the few people who need to be versed in the techniques and send them to a course on quality control. There is no need to subject everyone involved in the SPC implementation to the additional training. It will unnecessarily clutter the training program with information they will neither need nor use.

In conjunction with the lesson plan itself there are two highly recommended training aids. One is a bead box and the other is a Quincunx box. The bead box comes with a large number of various colored beads and a collection paddle of some sort. The beads are held in a receptacle in which you can vary the number and color of beads. When you dip the paddle into the receptacle you withdraw a random selection of the beads from the population of beads held in the bead receptacle. By varying the number and colors of beads in the box you can demonstrate changes in the population distribution. The Quincunx box, on the other hand, is an intriguing device for demonstrating various population distribution patterns. The box itself holds a large number of beads in a reserve on one end of the box. The box is thin to where it maintains a one bead depth of beads and has a glass front on one side to observe bead movement. When the box is stood on end the beads all pass through a single entry point and down through a maze of pins. As the beads randomly bounce off of the pins in their downward travel, they end up stacked in a visible distribution pattern at the bottom of the box. The box can be used to demonstrate normal distributions and skewed distributions; with movement of box controls the population can be made wider or narrower. People involved in SPC training programs say you can teach more statictical concepts in 30 minutes with these two visual aids than you can teach by lecturing for hours. These boxes are easy and fun to use and are thought provoking for both teachers and students.

Deciding how and where to do the training is pretty straight forward with just a few tips to follow. Training should be conducted in a quiet location removed from distractions and interruptions in so far as possible. That is pretty common sense. The only other point to remember on how to train is do not overwhelm your students. This statistics business is new to most of them, so give them time to absorb it. For each group to be trained break their training up into several sessions. An hour, no more than two hours, of training at a time is good. If you try for more than two hours to a session your audience will start to drift on you. Try for several days break between sessions and have some sample problems to be worked on in the interim. This will give ideas time to sink in before moving on to new concepts.

In general, whether you buy a training program or develop one of your own, your training program should be like your overall implementation program: paced and purposeful. Undue haste will confuse the program and the people.

FEEDBACK: A PROGRAM FOR CONTINUING IMPROVEMENT

The feedback loop in SPC has two focuses. One is feedback to the management of the program itself. The other is the feedback to management for decision making purposes.

The feedback to the management of the SPC program tells you how well your implementation effort met the needs of the job. Were the charts that you developed easy to use by workers and inspectors? Did the training program really teach the people what they needed to know and did they retain it? Are supervisors and managers receiving statistical information and are they using it? There will likely be a need for new training sessions for new people and refresher courses in charting and chart interpretation for those already trained. There must be an avenue for all those already trained in SPC to input suggestions for program improvements. In addition, the people responsible for implementing the SPC program must get out among the workers and talk to people to look for areas of program improvement. To keep an SPC training program current and applicable requires continuing attention and revisions when necessary.

The feedback into the management decision making loop will likely evolve almost on its own once an SPC program is in place and working. It is the getting it started that often stumps people. You can take a given production facility, train the managers, supervisors and workers, develop and give them the charts they need and have them all fired up about statistical methods, yet nothing happens because they do not know where to start.

Actually, the techniques for getting started should have been covered in the training program, but by the time the SPC program is ready to be kicked off on the production floor people may have forgotten. There are two simple tools to use to not only kick off the SPC effort but to use on a continuing basis to identify areas to concentrate on. They are the Pareto Chart and the Cause-and-Effect Diagram.

The whole idea behind the Pareto principle as it relates to process control is that although there may be many kinds of possible defects in a product, the majority of process or quality problems can be traced to just a few. According to the Pareto principle you will probably find 80% of your problems resulting from 10% or less of the possible influencing factors. Constructing a Pareto chart is nothing more than listing the kinds of process or quality problems a given production area suffers from and, using historical data, making hash marks for the number of occurences of each type of problem. It will quickly become apparent that while there may be many potential problems the majority of occurrences are with an isolated few. Quite logically, Pareto charting should be done on a periodic basis to focus statistical efforts on the problem areas where they will do the most good. (Pareto charting can also be used for selecting which areas of the company to implement the SPC program in next.)

Once problem candidates have been nailed down using the Pareto chart a cause-and-effect diagram can be brought into play to further isolate causes of the problems. The Pareto chart tells you what function to implement SPC on next, the cause-and-effect diagram helps you break that function down into its component parts to aid in troubleshooting.

Most troubleshooting is common sense logic (which may be in short supply). The cause-and-effect diagram is listing on paper or on a chart all of the steps that occur in a particular function. Every step, every material input, and every machine that a product passes through is listed in sequence. Each of these steps could be a particular "cause" of a problem. For each of these causes, each of these steps, you note what effect that cause has on the process.

The cause-and-effect diagram can be drawn to suit the user, but it is usually drawn as a time line with the steps, inputs, and machines listed in sequence down the line. The quality problems are then tracked back up the time line to their point of origin, to their cause. The cause is then studied for problem resolution.

A cause-and-effect diagram takes a little thought to construct so as not to neglect any potential causes, but the concept and mechanics are simple enough. For all its simplicity it is a powerful aid in locating the specific origin of defects and capability problems and can be adapted to any kind of process.

Another aid for troubleshooting needs to be mentioned. When a specific process has been identified as a source of problems and a candidate for SPC, to speed up the troubleshooting process the control limits on control charts can be moved in from $3\hat{\sigma}$ limits to $2\hat{\sigma}$ limits. As you recall $\pm 3\hat{\sigma}$ limits encompass 99.7% of a normal population, $\pm 2\hat{\sigma}$ limits include 95% of the population. While $2\hat{\sigma}$ limits may lead to some wasted effort in investigating false leads, if you have a process you know has problems it will make the charts more sensitive and you will be less likely to miss opportunities to locate problems.

With the Pareto chart and cause-and-effect diagram in use, mid-management's desire for improved productivity through SPC will provide the channels for feedback on process control and improvement. The Pareto chart can be used to identify areas for implementing statistical methods, and should be used within areas as a means of listing and isolating major origins of quality problems. Narrowed control limits and cause-and-effect diagrams can then be used to troubleshoot the problems. Throughout all this, upper management must remain active in the program to ensure its continuation and success and to monitor and direct quality improvement efforts. They will soon see the benefits of using SPC for the good of the business for short and long range planning and decision making.

COMPUTERIZATION POSSIBILITIES

Computers are marvelous tools for collecting, retaining and manipulating information. Statistical process data can be entered into a computer terminal at the work station in place of recording data on a chart. For manipulating data and performing calculations, if the information is accurately entered and the computer program is properly written the computer's calculations are virtually infallible. Computers can plot data, create graphs and charts, store historical data, check trends and runs, and they are tireless.

As to whether to computerize an SPC program, though, there are a number of factors that come in to play. First, most calculations in SPC are pretty simple and deal with just a few numbers at a time. Gross calculation errors are rather easy to recognize. Then again, recording and calculating is tedious and you can enter readings or observations on a computer keyboard as fast or faster than you can write them down. Once the data is in the computer results and graphs can be had at the touch of a key, with runs and trends and out of control conditions highlighted. A computer can even be programmed to recalculate and regraph your entire chart every time you input the data from a new sample; sort of a running graph, so to speak.

The big problem with computers is that they cost money. The more you want them to do the more money they cost. Depending on the size of the organization, you can have stand-alone microcomputers at work stations or you can have remote terminals at work stations tied into a larger computer. If remote terminals are used they can be simple entry key pads or full video terminals with graphic display chart capability. It all depends on what you want the computers to do. The computers are convenient, accurate and fast. The trade off is cost.

If you want to computerize an SPC program there is also the not so small problem of developing a computer program to fit the needs of the SPC program in question. As with SPC training programs, there are many off the shelf SPC computer programs. Unlike the training programs though, many of the computer programs are not near as flexible as most training programs.

Most computer programs were written with some specific application in mind and then attempts were made to make them more global in application. As a result, there are a lot of computer programs floating around that are advertised as SPC programs, yet they are limited in what they can do and are somewhat less than user friendly. Many claim great capabilities and carry a handsome price tag, but to make them serviceable for any specific application requires extensive user modification.

Do not let the problems of computerization dissuade you. If computerization promises to be a time saver and will improve accuracy and speed information transfer in your organization, fine. If you want to purchase an SPC program be cautious. Check the programs extensively for adequacy, expandability and vendor service. We come back to viewing SPC as another company program: if you were going to computerize another company program you would evaluate available computer programs on sufficiency and user friendliness, then you would compare available programs against what you could develop for yourself using either in-house or hired expertise.

SUMMARY

This chapter has looked at an SPC program from the point of reference of simply another company program, highlighting differences and problems specific to SPC. An SPC program, like most programs, is pushed based on its expected profitability. It is profitable. It reduces waste and improves quality. The constant tracking and quick identification of quality problems at all phases of production provides a constant means of monitoring the pulse of quality at all levels.

For more information on the constantly growing collection of computer software programs, see the "Annual QA/QC Software Directory." This is published annually in the March issue of *Quality Progress* and is an invaluable source of leads to currently available software.

The major difference that separates SPC from other new and yet routine company programs is the level of management involvement required. Upper level management must become and stay actively involved in the SPC program to ensure its success. The management information provided by the SPC system will enhance management's ability to make smart policy decisions. Increased productivity through reduced waste and improved quality are both short and long term benefits of statistical process control.

APPENDIX

Control Chart Constants and Formulas

Numbers in parentheses( ) indicate page numbers where formula explanations are found in the text.

1. Factors to use with $\bar{X}$ and R Control Charts

n	D_3	D_4	A_2	d_2
2	0	3.27	1.88	1.13
3	0	2.57	1.02	1.69
4	0	2.28	0.73	2.06
5	0	2.11	0.58	2.33
6	0	2.00	0.48	2.53
7	0.08	1.92	0.42	2.70
8	0.14	1.86	0.37	2.85
9	0.18	1.82	0.34	2.97
10	0.22	1.78	0.31	3.08

Factors to use with $\bar{X}$ and s Control Charts

n	A_3	B_3	B_4	c_4
2	2.66	*	3.27	.798
3	1.95	*	2.57	.886
4	1.63	*	2.27	.921
5	1.43	*	2.09	.940
6	1.29	.03	1.97	.952
7	1.18	.12	1.88	.959
8	1.10	.19	1.82	.965
9	1.03	.24	1.76	.969
10	.98	.28	1.72	.973

2. Formulas for $\bar{X}$ and R charts

$\bar{X}$ = the sum of all readings in a sample divided by the number of readings in the sample

R = the largest value in a sample minus the smallest value in the sample

$\bar{\bar{X}}$ = the sum of all sample means ($\bar{X}$'s) divided by the number of samples taken

$\bar{R}$ = the sum of all sample ranges (R's) divided by the number of samples taken

$\bar{X}$ Chart Control Limits (p. 21)

$$UCL = \bar{\bar{X}} + A_2\bar{R}$$

$$LCL = \bar{\bar{X}} - A_2\bar{R}$$

R Chart Control Limits $\qquad$ (p. 22)

$$UCL = D_4 \bar{R}$$

$$LCL = D_3 \bar{R}$$

$$\hat{\sigma} = \bar{R}/d_2 \qquad \text{(p. 11)}$$

Capability study formulas for use with $\bar{X}$ and R charts

For a more accurate $\hat{\sigma}$ for use in estimating $\pm 3\sigma$ population limits

$$\hat{\sigma} = \sqrt{\frac{n\Sigma x_i^2 - (\Sigma x_i)^2}{n(n-1)}} \qquad \text{(p. 27)}$$

or, for small sample sizes of 4 or 5 use the $\hat{\sigma}$ obtained from $\bar{R}/d_2$ and use $\pm 4\hat{\sigma}$ to estimate the population $\pm 3\sigma$ limits $\qquad$ (p. 27)

3. Formulas for use with s charts $\qquad$ (p. 39)

$$s = \sqrt{\frac{\Sigma x_i^2 - n\bar{X}^2}{n-1}}$$

$\bar{s}$ = Sum of all sample s's divided by number of samples.

$\bar{X}$ chart control limits using $\bar{s}$

$$UCL = \bar{\bar{X}} + A_3 \bar{s}$$
$$LCL = \bar{\bar{X}} - A_3 \bar{s}$$

s chart control limits

$$UCL = B_4 \bar{s}$$

$$LCL = B_3 \bar{s}$$

$$\hat{\sigma} = \bar{s}/c_4$$

4. Formulas for use with p charts $\qquad$ (p. 42 & 45)

$$\hat{\sigma} = \sqrt{\frac{\bar{p}(1 - \bar{p})}{\bar{n}}}$$

$$UCL = \bar{p} + 3\hat{\sigma}$$

$$LCL = \bar{p} - 3\hat{\sigma}$$

5. Formulas for use with np charts (p. 47)

$$\hat{\sigma} = \sqrt{\bar{d}\left(1 - \frac{\bar{d}}{\bar{n}}\right)}$$

$$\text{UCL} = \bar{d} + 3\hat{\sigma}$$

$$\text{LCL} = \bar{d} - 3\hat{\sigma}$$

6. Formulas for use with c charts (p. 47)

$$\hat{\sigma} = \sqrt{\bar{c}}$$

$$\text{UCL} = \bar{c} + 3\sqrt{\bar{c}}$$

$$\text{LCL} = \bar{c} - 3\sqrt{\bar{c}}$$

7. Formulas for use with u charts (p. 48)

$$\hat{\sigma} = \sqrt{\bar{u}/\bar{n}}$$

$$\text{UCL} = \bar{u} + 3\sqrt{\bar{u}/\bar{n}}$$

$$\text{LCL} = \bar{u} - 3\sqrt{\bar{u}/\bar{n}}$$

Z Chart for Standard Normal Distributions

Proportion of Total Area under the Curve to the Left of a Vertical Line Drawn at $\bar{X} + Z\sigma$, Where Z Represents Any Desired Value from $Z = 0$ to $Z = \pm 3.9$

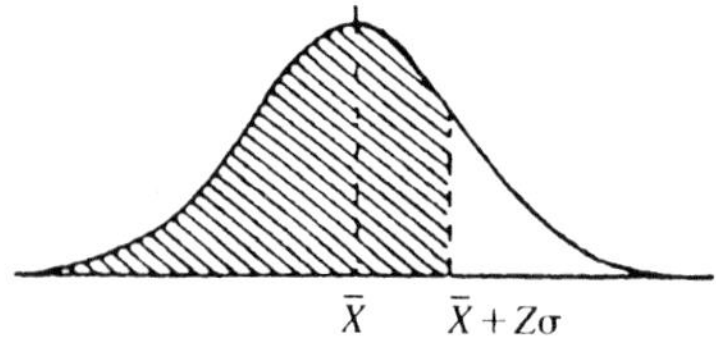

Z	0.09	0.08	0.07	0.06	0.05	0.04	0.03	0.02	0.01	0.00
—3.0	0.00100	0.00104	0.00107	0.00111	0.00114	0.00118	0.00122	0.00126	0.00131	0.00135
—2.9	0.0014	0.0014	0.0015	0.0015	0.0016	0.0016	0.0017	0.0017	0.0018	0.0019
—2.8	0.0019	0.0020	0.0021	0.0021	0.0022	0.0023	0.0023	0.0024	0.0025	0.0026
—2.7	0.0026	0.0027	0.0028	0.0029	0.0030	0.0031	0.0032	0.0033	0.0034	0.0035
—2.6	0.0036	0.0037	0.0038	0.0039	0.0040	0.0041	0.0043	0.0044	0.0045	0.0047
—2.5	0.0048	0.0049	0.0051	0.0052	0.0054	0.0055	0.0057	0.0059	0.0060	0.0062
—2.4	0.0064	0.0066	0.0068	0.0069	0.0071	0.0073	0.0075	0.0078	0.0080	0.0082
—2.3	0.0084	0.0087	0.0089	0.0091	0.0094	0.0096	0.0099	0.0102	0.0104	0.0107
—2.2	0.0110	0.0113	0.0116	0.0119	0.0122	0.0125	0.0129	0.0132	0.0136	0.0139
—2.1	0.0143	0.0146	0.0150	0.0154	0.0158	0.0162	0.0166	0.0170	0.0174	0.0179
—2.0	0.0183	0.0188	0.0192	0.0197	0.0202	0.0207	0.0212	0.0217	0.0222	0.0228
—1.9	0.0233	0.0239	0.0244	0.0250	0.0256	0.0262	0.0268	0.0274	0.0281	0.0287
—1.8	0.0294	0.0301	0.0307	0.0314	0.0322	0.0329	0.0336	0.0344	0.0351	0.0359
—1.7	0.0367	0.0375	0.0384	0.0392	0.0401	0.0409	0.0481	0.0427	0.0436	0.0446
—1.6	0.0455	0.0465	0.0475	0.0485	0.0495	0.0505	0.0516	0.0526	0.0537	0.0548
—1.5	0.0559	0.0571	0.0582	0.0594	0.0606	0.0618	0.0630	0.0643	0.0655	0.0668
—1.4	0.0681	0.0694	0.0708	0.0721	0.0735	0.0749	0.0764	0.0778	0.0793	0.0808
—1.3	0.0823	0.0838	0.0853	0.0869	0.0885	0.0901	0.0918	0.0934	0.0951	0.0968
—1.2	0.0985	0.1003	0.1020	0.1038	0.1057	0.1075	0.1093	0.1112	0.1131	0.1151
—1.1	0.1170	0.1190	0.1210	0.1230	0.1251	0.1271	0.1292	0.1314	0.1335	0.1357
—1.0	0.1379	0.1401	0.1423	0.1446	0.1469	0.1492	0.1515	0.1539	0.1562	0.1587
—0.9	0.1611	0.1635	0.1660	0.1685	0.1711	0.1736	0.1762	0.1788	0.1814	0.1841
—0.8	0.1867	0.1894	0.1922	0.1949	0.1977	0.2005	0.2033	0.2061	0.2090	0.2119
—0.7	0.2148	0.2177	0.2207	0.2236	0.2266	0.2297	0.2327	0.2358	0.2389	0.2420
—0.6	0.2451	0.2483	0.2514	0.2546	0.2578	0.2611	0.2643	0.2676	0.2709	0.2743
—0.5	0.2776	0.2810	0.2843	0.2877	0.2912	0.2946	0.2981	0.3015	0.3050	0.3085
—0.4	0.3121	0.3156	0.3192	0.3228	0.3264	0.3300	0.3336	0.3372	0.3409	0.3446
—0.3	0.3483	0.3520	0.3557	0.3594	0.3632	0.3669	0.3707	0.3745	0.3783	0.3821
—0.2	0.3859	0.3897	0.3936	0.3974	0.4013	0.4052	0.4090	0.4129	0.4168	0.4207
—0.1	0.4247	0.4286	0.4325	0.4364	0.4404	0.4443	0.4483	0.4522	0.4562	0.4602
—0.0	0.4641	0.4681	0.4721	0.4761	0.4801	0.4840	0.4880	0.4920	0.4960	0.5000

Z	0.00	0.01	0.02	0.03	0.04	0.05	0.06	0.07	0.08	0.09
+0.0	0.5000	0.5040	0.5080	0.5120	0.5160	0.5199	0.5239	0.5279	0.5319	0.5359
+0.1	0.5398	0.5438	0.5478	0.5517	0.5557	0.5596	0.5636	0.5675	0.5714	0.5753
+0.2	0.5793	0.5832	0.5871	0.5910	0.5948	0.5987	0.6026	0.6064	0.6103	0.6141
+0.3	0.6179	0.6217	0.6255	0.6293	0.6331	0.6368	0.6406	0.6443	0.6480	0.6517
+0.4	0.6554	0.6591	0.6628	0.6664	0.6700	0.6736	0.6772	0.6808	0.6844	0.6879
+0.5	0.6915	0.6950	0.6985	0.7019	0.7054	0.7088	0.7123	0.7157	0.7190	0.7224
+0.6	0.7257	0.7291	0.7324	0.7357	0.7389	0.7422	0.7454	0.7486	0.7517	0.7549
+0.7	0.7580	0.7611	0.7642	0.7673	0.7704	0.7734	0.7764	0.7794	0.7823	0.7852
+0.8	0.7881	0.7910	0.7939	0.7967	0.7995	0.8023	0.8051	0.8079	0.8106	0.8133
+0.9	0.8159	0.8186	0.8212	0.8238	0.8264	0.8289	0.8315	0.8340	0.8365	0.8389
+1.0	0.8413	0.8438	0.8461	0.8485	0.8508	0.8531	0.8554	0.8577	0.8599	0.8621
+1.1	0.8643	0.8665	0.8686	0.8708	0.8729	0.8749	0.8770	0.8790	0.8810	0.8830
+1.2	0.8849	0.8869	0.8888	0.8907	0.8925	0.8944	0.8962	0.8980	0.8997	0.9015
+1.3	0.9032	0.9049	0.9066	0.9082	0.9099	0.9115	0.9131	0.9147	0.9162	0.9177
+1.4	0.9192	0.9207	0.9222	0.9236	0.9251	0.9265	0.9279	0.9292	0.9306	0.9319
+1.5	0.9332	0.9345	0.9357	0.9370	0.9382	0.9394	0.9406	0.9418	0.9429	0.9441
+1.6	0.9452	0.9463	0.9474	0.9484	0.9495	0.9505	0.9515	0.9525	0.9535	0.9545
+1.7	0.9554	0.9564	0.9573	0.9582	0.9591	0.9599	0.9608	0.9616	0.9625	0.9633
+1.8	0.9641	0.9649	0.9656	0.9664	0.9671	0.9678	0.9686	0.9693	0.9699	0.9706
+1.9	0.9713	0.9719	0.9726	0.9732	0.9738	0.9744	0.9750	0.9756	0.9761	0.9767
+2.0	0.9773	0.9778	0.9783	0.9788	0.9793	0.9798	0.9803	0.9808	0.9812	0.9817
+2.1	0.9821	0.9826	0.9830	0.9834	0.9838	0.9842	0.9846	0.9850	0.9854	0.9857
+2.2	0.9861	0.9864	0.9868	0.9871	0.9875	0.9878	0.9881	0.9884	0.9887	0.9890
+2.3	0.9893	0.9896	0.9898	0.9901	0.9904	0.9906	0.9909	0.9911	0.9913	0.9916
+2.4	0.9918	0.9920	0.9922	0.9925	0.9927	0.9929	0.9931	0.9932	0.9934	0.9936
+2.5	0.9938	0.9940	0.9941	0.9943	0.9945	0.9946	0.9948	0.9949	0.9951	0.9952
+2.6	0.9953	0.9955	0.9956	0.9957	0.9959	0.9960	0.9961	0.9962	0.9963	0.9964
+2.7	0.9965	0.9966	0.9967	0.9968	0.9969	0.9970	0.9971	0.9972	0.9973	0.9974
+2.8	0.9974	0.9975	0.9976	0.9977	0.9977	0.9978	0.9979	0.9979	0.9980	0.9981
+2.9	0.9981	0.9982	0.9983	0.9983	0.9984	0.9984	0.9985	0.9985	0.9986	0.9986
+3.0	0.99865	0.99869	0.99874	0.99878	0.99882	0.99886	0.99889	0.99893	0.99896	0.99900

SOURCE: Grant and Leavenworth, Statistical Quality Control, 4 ed., 1972, pp. 642-43.
Reprinted from E. R. Ott, *Process Quality Control.* pp. 318-319

GLOSSARY OF TERMS

In an effort to keep terms standardized in the quality control and statistical process control fields, the following definitions were extracted from *Glossary and Tables for Statistical Quality Control,* prepared by and available through the American Society for Quality Control, Statistics Division.

Acceptance (control chart or acceptance control chart usage). A decision that the process is operating in a satisfactory manner with respect to the statistical measures being plotted.

Accuracy. The closeness of agreement between an observed value and an accepted reference value.

Arithmetic mean and arithmetic average. A measure of central tendency or location which is the sum of the observations divided by the number of observations.

Assignable cause. A factor which contributes to variation and which is feasible to detect and identify.

Attributes, Method of. Measurement of quality by the method of attributes consists of noting the presence (or absence) of some characteristic or attribute in each of the units in the group under consideration, and counting how many units do (or do not) possess the quality attribute, or how many such events occur in the unit, group or area.

c (count). The count or number of events of a given classification occuring in a sample. More than one event may occur in a unit (area of opportunity), and each such event throughout the sample is counted.

Chance causes. Factors, generally numerous and individually of relatively small importance, which contribute to variation, but which are not feasible to detect or identify.

Chance variation. Variation due to chance causes.

Characteristic. A property which helps to differentiate between items of a given sample or population.

Control chart. A graphical method for evaluating whether a process is or is not in a "state of statistical control" (also known as Shewhart Control Chart).

Control chart factor. A factor, usually varying with sample size, to convert specified statistics or parameters into a central line value or control limit appropriate to the control chart.

Control chart method. The method of using control charts to determine whether or not processes are in a stable state.

Control limits. Limits on a control chart which are used as criteria for signaling the need for action, or for judging whether a set of data does or does not indicate a "state of statistical control". Lower control limits are for points plotting below the center level. Upper control limits are for points plotting above the center level.

Defect. A departure of a quality characteristic from its intended level or state that occurs with a severity sufficient to cause an associated product or service not to satisfy intended normal, or reasonably foreseeable, usage requirements.

Defective unit. A unit of product or service containing at least one defect, or having several imperfections that in combination cause the unit not to satisfy intended normal, or reasonably foreseeable, usage requirements.

Deviation. The difference between a measurement or quasi-measurement and its stated value or intended level.

Event. An occurrence of some attribute.

Hypothesis. a). Null hypothesis — The hypothesis tested in tests of significance is that there is no difference (null) between the population of the sample and the specified population (or between the populations associated with each sample). The null hypothesis can never be proved true. It can, however, be shown, with specified risks of error, to be untrue; that is, a difference can be shown to exist between the populations. If it is not disproved, one usually acts on the assumption that there is no adequate reason to doubt that it is true. b). Alternative hypothesis — The hypothesis that is accepted if the null hypothesis is disproved.

In-Control process. A process in which the statistical measure(s) being evaluated are in a "state of statistical control".

Inspection. The process of measuring, examining, testing, gaging or otherwise comparing the unit with the applicable requirements.

Item. An object or quantity of material on which a set of observations can be made.

Lower control limit (LCL). see Control Limits

Median. For an ordered set of numbers, the middle measurement when an odd number of units are arranged in order of size. When an even number of units are so arranged, the median is the average of the two middle units.

Mode. The most frequent value of the variable.

Natural process limits. Limits which include a stated fraction of the individuals in a population. For populations with a Normal (Gaussian) Distribution, the natural process limits ordinarily will be set at ± 3. If placed around the standard level, these limits identify the boundaries which will include 99.7% of the individuals in a process that is properly centered and in a state of statistical control.

Nonconformity. A departure of a quality characteristic from its intended level or state that occurs with a severity sufficient to cause an associated product or service not to meet a specification requirement.

np (number of affected units). The total number of units (areas of opportunity) in a sample in which an event of a given classification occurs. A unit (area of opportunity) is to be counted only once even if several events of the same classification are encountered therein.

Nonconforming unit. A unit of product or service containing at least one nonconformity.

Observed value. The particular value of a characteristic determined as a result of a test or measurement.

p (percent of units affected). The percentage of the total number of units (areas of opportunity) in a sample in which an event of a given classification occurs. A unit (area of opportunity) is to be counted only once even if several events of the same classification are encountered therein.

Population. The totality of items or units of material under consideration.

Process capability. The limits within which a tool or process operates based upon minimum variability as governed by the prevailing circumstances.

Quality. The totality of features and characteristics of a product or service that bear on its ability to satisfy given needs.

R (range). A measure of dispersion which is the difference between the largest observed value and the smallest observed value in a given sample.

$\overline{R}$. The average for the set under consideration of sample values of R.

Random causes. see Chance causes.

Random sampling. The process of selecting units for a sample of size n in such a manner that all combinations of n units under consideration have an equal or ascertainable chance of being selected as the sample.

Random variation. see Chance variation

Run. An uninterrupted sequence of occurrences of the same attribute or event in a series of observations, or a consecutive set of successively increasing (run "up") or successively decreasing (run "down") values in a series of variable measurements.

s (sample standard deviation). A measure of variability (dispersion) of observations in the sample that is the positive square root of the sample variance.

$\bar{s}$. The average for the set under consideration of sample values of s.

Sample. A group of units, portion of material, or observations taken from a larger collection of units, quantity of material, or observations that serves to provide information that may be used as a basis for making a decision concerning the larger quantity.

Sample size. The number of units in a sample.

Sampling interval. In systematic sampling, the fixed interval of time, output, running hours, etc. between samples.

Specification limits. see Tolerance Limits

Standard deviation. A measure of variability (dispersion) of observations that is the positive square root of the population variance.

State of statistical control. A process is considered to be in a "state of statistical control" if the variations among the observed sampling results from it can be attributed to a constant system of chance causes.

Statistic. A quantity calculated from a sample of observations, most often to form an estimate of some population parameter.

Statistical measure. A statistic or mathematical function of a statistic.

Tolerance (Specification). The total allowable variation around a level or state (upper limit minus lower limit), or the maximum acceptable excursion of a characteristic.

Tolerance limits (Specification limits). Limits that define the conformance boundaries for an individual unit of a manufacturing or service operation.

u (count per unit). The average count, or average number of events of a given classification, per unit (unit area of opportunity) occurring within a sample. More than one event may occur in a unit (unit area of opportunity), and each such event is counted.

Unit. A quantity of product, material or service forming a cohesive entity on which a measurement or observation may be made.

Upper control limit (UCL). see Control Limits

Variables, method of. Measurement of quality by the method of variables consists of measuring and recording the numerical magnitude of a quality characteristic for each of the units in the group under consideration. This involves reference to a continuous scale of some kind.

Variance. A measure of variability (dispersion) of observations based upon the mean of the squared deviations from the arithmetic mean.

X. Sample average.

$\bar{X}$. The average for the set under consideration of sample values of X.

REFERENCE SOURCES

Articles

Roger L. Berger. "Looking for Software." Directory of Software for QA or QC, *Quality Progress,* Mar 1984, pp. 28-30

Paul C. Clifford. "A Process Capability Study Using Control Charts." *Journal of Quality Technology,* July 1971, pp. 107-111

Paul C. Clifford. "Control Charts Without Calculations." *Industrial Quality Control,* May 1959, pp. 2-6

W. Edwards Deming. "Improvement of Quality and Productivity through Action by Management." *National Productivity Review,* Winter 1981-82, pp. 12-22

W. Edwards Deming. "On Some Statistical Aids Toward Economic Production." *Interfaces,* Aug 1975, pp. 1-15

Gary L. Earnhart. *Statistical Control of Suspension Bushing Assembly.* SAE Paper 830300, 1983, pp. 39-42

Perry Gluckman. *Teaching the Deming Statistical Quality Control Method.* SAE Paper 820980, 1982, pp. 75-79

Jeremy Main. "Ford's Drive for Quality." *Fortune,* 18 Apr 1983, pp. 62-70

Hobart Rowe. In Letters to the Editor, *Wall Street Journal,* 9 Mar 1983

James C. Siegel. *Managing with Statistical Methods.* SAE Technical Paper Series #820520, 1982

James C. Siegel. " Statistical Management Methods to Improve Quality, Productivity and Competitive Position." Industrial Liason Program Symposium, Massachusetts Institute of Technology, 17 Aug 1982

Andrew L. Strongrich, Gerald E. Herbert, and Thomas J. Jacoby. *Simple Statistical Methods,* SAE Paper 830299, 1983, pp. 31-38.

Thomas R. Temin. "Ford to Suppliers: Get Serious About Quality." *Purchasing Magazine,* 27 Feb 1983

"Statistical Quality Control." Special Report #762, *American Machinist.* Jan 1984, pp. 97-108

Books

W. Edwards Deming. *Quality, Productivity and Competitive Position.* Massachusetts Institute of Technology, 1982

A. V. Feigenbaum. *Total Quality Control,* 3d ed. New York: McGraw-Hill Co., 1983

Eugene L. Grant, Richard S. Leavenworth. *Statistical Quality Control.* New York: McGraw-Hill Co., 1972

J. M. Juran, editor-in-chief, Frank M Gryna, Jr., associate editor and R. S. Bingham, Jr., associate editor. *Quality Control Handbook,* 3d ed. New York: McGraw-Hill Co., 1974

Ellis R. Ott. *Process Quality Control.* New York: McGraw-Hill Co., 1975

Capability Study Guidelines. Deere & Company, Reliability Department, Aug 1981

Continuing Process Control and Process Capability Improvement: A Guide to the Use of Control Charts for Improving Quality and Productivity for Company, Supplier and Dealer Activities. Statistical Methods Office, Operations Support Staffs, Ford Motor Company, 1983

Statistical Process Control. Technical Training, United Technologies, Pratt & Whitney

In addition to the above listed articles and books, grateful appreciation is due to Mr. David R. Schwinn of the Product Quality Policy and Planning Department, Product Quality Office, Operations Support Staffs, Ford Motor Company. Mr. Schwinn provided numerous case studies used in the Ford Motor Company training programs which were invaluable in compiling this manual.

INDEX